The Wishing Tree

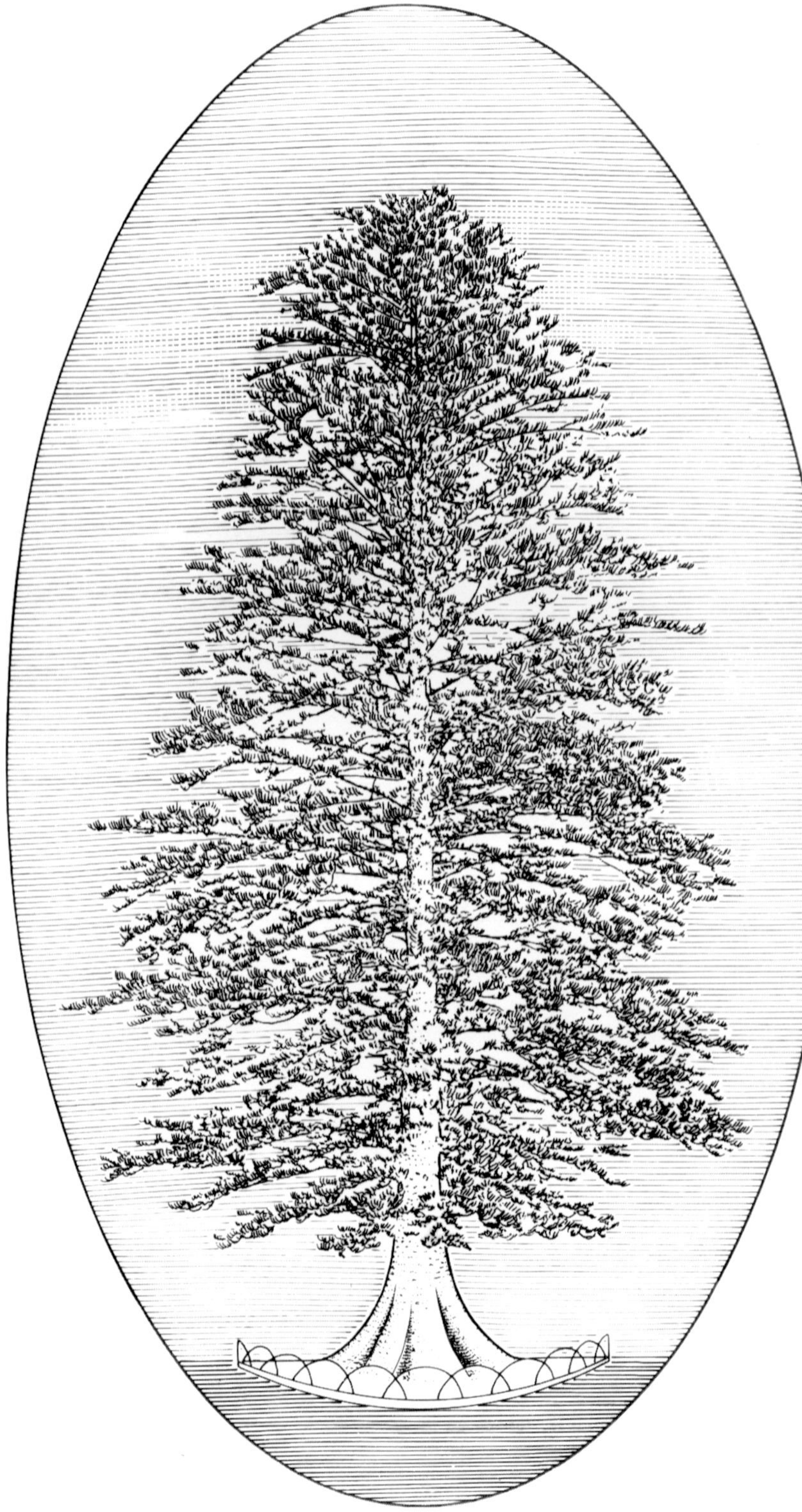

The Wishing Tree

A Guide to memorial trees, statues, fountains, etc. in the Royal Botanic Gardens, Domain, and Centennial Park, Sydney

Edwin Wilson

Kangaroo Press

in association with the

Royal Botanic Gardens Sydney

BP Australia is very pleased to continue its involvement with the historic Royal Botanic Gardens by supporting this book. As part of its bicentenary contributions, BP sponsored the interpretative display at the Botanic Gardens' Tropical Centre which was opened in 1990 by the Governor of New South Wales.

We compliment the author, Edwin Wilson, for contributing this worthy addition to Australia's heritage.

The Council of the City of Sydney is proud to be associated with the publication of *The Wishing Tree.* A Sesquicentenary Endorsed Project.

Jacket design by Darian Causby using a coloured illustration from *The Picturesque Atlas of Australasia* by kind permission of David Bedford and a drawing of the Wishing Tree by David Mackay.

First published in 1992 by Kangaroo Press Pty Ltd
3 Whitehall Road (P.O. Box 75) Kenthurst NSW 2156
Typeset by G.T. Setters Pty Limited
Printed in Singapore through Global Com Pte Ltd

ISBN 0 86417 4373

Contents

Acknowledgments

In 1968 I attended a series of lectures by a colleague, Dr Lionel A. Gilbert, at what was then Armidale Teachers' College, on the subject of Australian history. These lectures inspired me to seek out my own roots, and in researching my family history over the subsequent years I became much more aware of aspects of Australian history.

A chance meeting and subsequent friendship with the Polish sculptor, Lucian Bruno Michalski, opened windows to old Europe, classicism, modernism, and the art of sculpture. Then when I worked at the Australian Museum I coordinated the research and production of an animated film entitled *The First 150 Years*, on the history of the Museum for its sesquicentennial celebrations. The film incorporated the history of the Museum with Australian social history, and the history of science. And as the Museum and the Gardens had the same trustees in the early years, this served as an introduction to some of the players and issues involved.

When I came to work at the Gardens in 1980 I inherited the initial file accumulated by Betty Jacobs, the first Education Officer — as published in her booklet, *Statues and Memorials* — who had handed the torch on to Diane Alford, who handed it on to me. From the beginning of my employment here it was apparent that there was considerable public interest in the statues and fountains, gates, buildings and memorial trees. I established my own ever-expanding file on the subject, with the intention of turning it into a publication at some future time, but continually put it off because it wasn't 'finished'.

It was the requirement to compile a heritage listing of statues and memorials for a Heritage and Conservation Register, in late 1988 (according to the guidelines of The Heritage Act, as amended in 1987) that finally precipitated this work. The Bicentennial year (with a few exceptions) has been taken as an arbitrary point at which to cut off this compilation.

Such projects are open-ended almost by definition and will never be finished. I now believe there is sufficient information to make for interesting reading, but I would welcome any additional material from readers.

The need to research the history of the Gardens for Chapter 1 of the 'Blue Book' or 'Guide to the Gardens', *Royal Botanic Gardens Sydney* (published in 1982) really threw me in at the deep end.

Primary material was obtained from the State Library of New South Wales, the State Archives, the collection of glass negatives at The N.S.W. Government Printer, from Harry Craig (grandson of former Director, J. H. Maiden), from annual reports and the excellent collection of old photographs in the Gardens' Library. Anna Hallett, Gardens' Librarian, has always been most helpful, and Irma Trunov, Library Assistant, translated the inscriptions on the Sebastopol cannons in Centennial Park. I would like to record an appreciation for the work of Jill Gaynor, who was working for me as an assistant in 1980, and who spent a considerable amount of her time in this early period at the State Library and the State Archives.

Being so close to the State Library of New South Wales, with its Mitchell and Dixson wings, made it convenient to check through the various references in their picture catalogues and indexes. In this context I would specially like to mention Mr Con Davis, Title Maps Cataloguer, who first showed me the original 1816 Cartwright Map of the Gardens and Domain.

Secondary sources included various published papers by R. H. Cambage and Walter F. Froggatt from the papers of the late J. H. Maiden; the *Guide to the Gardens* by J. H. Maiden (1903); An *ABC of the Royal Botanic Gardens, Sydney*, by R. H. Anderson; the unpublished University of New England (1971) PhD thesis of Lionel Gilbert, *Botanical Investigations of New South Wales, 1811-1880*, (Chapter 5 of which concerned itself with the Sydney Botanic Gardens); *The Royal Botanic Gardens, Sydney: A History 1816-1985*, by Lionel Gilbert, published by Oxford University Press in 1986; and David Churches' manuscript, *The Cultural Significance of the Royal Botanic Gardens, Sydney* (1987).

Books, like exhibitions, move in mysterious ways. Some time in 1982, after having written the chapter on the history of the Gardens for the 'Blue Book', I was rung by Joan Simms, of the Premier's Department, who asked me if I had any information on 'the plaque near the ladies toilet block in the Domain'. Not being one who normally frequents ladies' toilets, I said I didn't have a clue. But being a good public servant I said I'd try to find out, and this marked the beginning of my interest in the history of the Domain.

Then, in 1984, Shirley Colless was employed on Trust funds to work with me as a research assistant for the production of an exhibition entitled 'Discovering the Domain' and I am indebted to her original research, later shared with David Churches, and to Michelle Frank who assisted with the bulk of the exhibition text. The book derived from this exhibition, *Discovering the Domain*, a pictorial social history of the Sydney Domain, was published in 1986 by Hale & Iremonger.

Since work on this compilation started at the end of 1988, I have also been indebted to Heather Edwards, who spent hours in the Gardens and Domain, measuring and transcribing the information on plaques and statues; and to Tony Jeans, who initially worked as a volunteer in the capacity of a 'Girl Friday' (and more recently on Trust funds), and has spent many hours in the State Library assiduously following up leads and checking details of dates, footnotes, photographs, etc. Her continuity in this respect has been much appreciated in a sea of interruptions.

Then, when the manuscript was being compiled, I was talking to ecologist Doug Benson, at morning tea. By some strange serendipity he had a cutting (believed to be from the *Sydney Morning Herald*) that identified the source of eight statues in the Gardens, including the Four Seasons. Usually it was never as easy as this. Unfortunately the source and date of the article had not been recorded. Tony Jeans was given this job, working back from 22 September 1882, as it had been recorded that one of the Seasons at the top of the stairs had been damaged by falling debris during the conflagration of the Garden Palace. When this search proved to be fruitless, detective work on other articles on the same page limited the dates to a 'window' between an inquest in Grafton New South Wales into the sinking of a certain ship, and when the railway reached Armidale, New South Wales, confirming the particular statues of the Seasons (as described) came to the Gardens after the fire.

A letter from Mr Mervyn Doust (following a newspaper article in the *Sydney Morning Herald* [27 April 1988] regarding the placement of some restored statues on the steps to the Palace Garden) confirmed the source of the other statue of Spring, which once stood near what is now Lion Gate Lodge. Then, following a newspaper report of later vandalism of that same statue of Spring, I received a letter from Judith Hunter, retired librarian, who reported a potential source of the original statues of the Seasons — which helped the scales fall from my eyes.

When the manuscript was in an advanced form I used David Churches' excellent thesis *The Cultural Significance of the Royal Botanic Gardens, Sydney*, for cross-referencing — and specifically for his invaluable research relating to buildings.

Draft chapters of this manuscript were typed by Desley Williams and Jean Westropp. The final manuscript was typed by Angela Benn and reviewed in a penultimate form by Kathleen Wall, friend and Volunteer Guide at the Gardens. Her wise counsel was, as always, much appreciated.

The book was edited by Carl Harrison-Ford.
Layout and design was by Stephen Dearnley.
Photographs by Jaime Plaza.

Other Books by Edwin Wilson

Royal Botanic Gardens Sydney (ed.), Royal Botanic Gardens and Domain Trust, 1982

Banyan, a selection of poems with pencil drawings by Elizabeth McAlpine, Woodbine Press, 1982

Liberty, Egality, Fraternity!, a novel, Woodbine Press, 1984

The Dragon Tree, a selection of poems with pencil drawings by Elizabeth McAlpine, Woodbine Press, 1985

Discovering the Domain (ed.), a pictorial social history of the Sydney Domain, Hale & Iremonger, 1986

Wild Tamarind, a bicentennial parable and morality tale, a novel, Woodbine Press, 1987

Falling up into Verse (a Handbook [Survival Manual] for Live Poets, Aspiring Poets, and Students of Poetry, Down Under), Woodbine Press, 1989

Songs of the Forest, rainforest poems with pencil drawings by Elizabeth McAlpine, Hale & Iremonger, 1990

The Rose Garden, a selection of poems with pencil drawings by Elizabeth McAlpine, Woodbine Press, 1991

Dedicated to Tony (Antoinette) Jeans
for many hours spent following up leads
and footnotes, and for her continuity in
a sea of interruptions.

Introduction

The beautiful Royal Botanic Gardens and Domain, by the city and its harbour, have been visited by generations of children, students, lovers, tourists and family groups — on Sunday afternoons and during school holidays — with all the associated happy memories of picnics and promenades, ice-creams, ducks and statues. For many years the old Wishing Tree (Middle Garden, see p.49) was a focal point in these Gardens — a place for a secret wish or lover's tryst — and a link with the colonial past. Being so close to the heart of the city the Gardens and Domain have been silent witnesses to the picket fence and the parade of European history in Australia, from the first farm to the present day.

The great virtue of public gardens lies in their continuity. As such the Botanic Gardens reflect the various historical moods — from the utilitarian rectangular beds of the monastic tradition of the Middle Garden ('short back and sides'); the sweeping pathways of 'Capability' Brown's landscaping style of the Lower Garden with its vistas and panoramas and water features; the 1960s 'patio' style (albeit vertical!) of the Succulent Garden; the post-Whitlamesque 'native' garden of the 1970s; the formal and classical beauty of the New Rose Garden (1987); to the stainless-steel neo-Victorian revivalist 'conservatory feel' of the new Tropical Centre — and are close to the heart of the people.

I fell in love with these Gardens when I first came to Sydney in 1962, and was a frequent visitor — as a tourist, and later as a student of botany — and I even tried my hand at sketching some statues. For some, the associations run deeper, as the Gardens have been inexorably linked with the rites of passage of their very lives. Affairs of the heart and business have been instigated here, and ended; marriages sealed and celebrated; divorce settlements negotiated on garden seats; and even drug deals consummated. Individuals have been conceived and died here, of natural or other causes, in an environment of beauty and peace.

The Gardens are an oasis of tranquillity and beauty. But do not be deceived by the superficial calm, for they were created through destruction and conflict — as various directors struggled to impose their wills upon the meagre soil of Woccanmagully (later called Farm Cove to commemorate another great Australian failure).

In fact, the Royal Botanic Gardens, Sydney, were born almost of sorrow, and forged through conflict — between Europeans and Aborigines; convicts and trees (John White wrote in his diary 'It will scarcely be credited when I declare that I have known twelve men employed for five days in grubbing out one tree'. Public service gardening indeed!); convict chain gangs and their supervisors; gardeners and the elements (poor soil, drought, disease); supervisors and their committees of management; directors and pinch-penny governments; land-grabbers, politicians with edifice complexes and those who would wish to preserve what is left — and finally between the creators themselves, in that one long clash of egos on this site for more than 200 years.

No such deliberations would be complete without a mention of the thousands of men (more recently men and women) who have carried dung and soil and water, and cut the grass and stone, and turned the sod and tilled the earth, and nurtured all the plantings over time. Generations of gardeners have lived on and/or worked this site since the hungry days when the farm failed. The first gangs served out their time at the pleasure of King George III, although some individuals, like the Scottish martyr Joseph Gerrald (see p.127), were illustrious tenants.

The current crop of gardeners are still paid 'convict' wages, and in this context it is interesting to note that the term 'supervisor' is still used to describe the person who controls the outdoor staff. The lucky ones, however, received cheap housing at a very choice address up to the 1990s. At least the contemporary gardener has the choice and many establish long and happy associations with the Gardens, through the nature of their work, at this most congenial of locations, and through the networks of their families and their friends.

Horace observed in one of his odes that when we launch a poem or plant a tree we lay claim to our own small share of immortality — and that we do not wholly die. Buildings, statues and fountains serve a similar function, and exist as memorials to individuals and their cosmologies. Many people must agree, for hardly a month goes by without us receiving a request to plant a tree for someone or other or some good cause, or erect yet another statue or memorial in the Gardens or Domain.

The files abound with requests for memorials, not forgetting the fictional statue of Johann Ulrich Voss erected in the Domain in Patrick White's *Voss.*[1] The character of Voss was modelled on Ludwig Leichhardt, who unsuccessfully applied for the position of director of the Sydney Gardens in 1842 and 1844,[2] and who perished in the desert while trying to cross the continent after heading west from the Condamine River, in 1848, at the age of thirty-four. In 1961 there was a proposal to erect a statue to the Tolpuddle Martyrs in the Domain.[3] Then several years ago an architect wanted to erect a giant water wheel between Farm Cove and the ponds in the Lower Garden. Mrs Macquaries Point has always been a prime target. In recent times we have had requests that a cannon, a cairn of stones from Scotland, and a sculpture of a nude female sitting on a rock (and a man beside her skimming stones across the water), be erected there!

The main problem is that the Gardens have limited space, so tree-planting — not to mention memorialising — has traditionally been restricted to royalty and politicians, with few exceptions.

At the dawn of civilization, Neolithic man arranged great stones in religious, ritualistic, astronomical and symbolic ways — as if to focus power, or draw it from the very earth — culminating in such masterpieces as Stonehenge. This almost innate reverence for stone surfaced in such disjunct forms as the Olmec heads of Central America, the stone statues of Easter Island, and carved wooden statues from around the world — making sculpture a very ancient art. The poem 'Ozymandias', by Shelley, evokes this mood:

> I met a traveller from an antique land
> Who said: Two vast and trunkless legs of stone
> Stand in the desert ... Near them, on the sand,
> Half sunk, a shattered visage lies, whose frown,
> And wrinkled lip, and sneer of cold command,
> Tell that its sculptor will those passions read ...[4]

Desmond Kingthele in *Shelley: His Life and Work*, refers to the 'traveller', Dr Pococke, who described such statues as Ozymandias (Ramases II). The inscription on the Egyptian tomb of Diodorus Suculus reads 'I am King of Kings', which was probably the inspiration for the poem.

The ancient Egyptians erected great obelisks and other monuments to their warrior kings and gods, the Pharaohs, who spent their lives constructing mausoleums. The Greeks later perfected representation of the idealised human form in metal and stone, as free-standing sculpture or as part of the friezes in buildings, while the Romans extended this concept to erect sculptures to their soldiers, politicians and gods as an integral part of town planning — at the focal points of roads, squares and buildings — and at points of intersection of pathways in parks and gardens.

Until recent times, most of our sculptural forms — columns, statues on pedestals, equestrian groups, and monumental arches — were derived from Greek or Roman models, often via the Renaissance, when interest in antiquity was revived and formal gardens decorated with neo-classical statues became popular again. Such gardens proliferated in Victorian times.

Because the bulk of the Sydney Gardens landscaping, especially the Lower Garden and Palace Garden, was laid out in the second half of the last century they have a distinctive Victorian character. Many of the important statues in the Gardens were also from this era and helped create the mood, while the garden pavilions (devised from earlier Persian models) have the feel of Edwardian 'summer houses'. With characteristic overkill the Victorians cluttered their gardens with a rash of statues, urns, carpet bedding, bells, beads and baubles in every grotto, nook, cranny and corner — and at the intersection of every major pathway. Old photos of these Gardens show they were no exception. Compared with the Palace Garden, most other areas were quite restrained, although a 1914 photograph shows that the main path through the Upper Garden, from the entrance gates south of the Wishing Tree, was also lined with four pairs of sculptures.

The Palace Garden was originally a flamboyant showpiece of classical statuary and garden ornaments, the majority of which were marble copies of classical originals imported for the Sydney International Exhibition in 1879. Some sculptures were installed in the Gardens as part of their beautification leading up to the Exhibition, others were part of external displays, and others may have been introduced when the Palace Garden was being developed around 1884.[5]

Sculpture had been snubbed at the earlier Intercolonial Exhibition of 1870, when the very word had been omitted from the prospectus, although some eight works (including busts and medallions) of Mr William Lorando Jones had been on display.[6] The situation was rectified for the International Exhibition. However, the main collection of statues (including Linceus and Hypermenestra which arrived in Sydney in 1875 by Charles Summers, father of Charles Francis Summers,[7] and a bronze of the young Queen were destroyed in the 1882 fire.[8]

Sir Henry Parkes was an enthusiastic promoter of public sculpture to beautify the colony with scaled-up and scaled-down copies from classical antiquity and the Victorian classicists. In the 1880s he was responsible for many such marbles, shipped especially from Italy for that purpose.[9]

And so, when our civic fathers placed imported statues in the gardens, they set the agenda of — 'Gods in the Gardens, nymphs in the bush'[10] — and the occasional polly (and later even poets) in the parks.

The original statues of the Four Seasons were imported from Italy in 1879 by Mr Job Hanson for the International Exhibition, and displayed on the Palace Garden steps along with two monuments (gravestones, see p.102).[11] In a letter to Sir Henry Parkes, 26 August 1881, Job Hanson writes from Carrara, Italy:

> I arr'd in Italy 10 days ago & have been busily employed seeing what could be had in the way of garden statuary . . . I have bought 8 statues as follows all really good things 'Charity' 5 ft, 'Venis' [*sic*] by Cannova [*sic*] 5'6'', 'Literature', 'Mechanica', 'Navigation', 'Astronomy', each 5'2'' high & really good works 'History' 5'6'', [and] 'Andromeda', a very nice work . . .
>
> I have made offers for 4 others & given commissions for the two like enclosed photos 'Rebecca', 'Bather' each 5'10'' high from a first class sculptor.[12]

Unfortunately Hanson does not give the name of the sculptor, but he goes on to say the statues from Carrara were as good as any work from Florence or Rome, and cheaper:

> I can get what others statues are required modelled from the best works by one of the best artists in Italy, & then made by good sculptors here, all say 5'10'' high or 6 feet . . .
>
> I leave for Milan to-morrow where an exhibition is being held. Something may be bought there, or ideas gained. From there I go to Paris where I believe the best garden statuary in the world is erected. I shall obtain photos of those I think suitable with a view to their being modelled after receiving final instructions from you by wire on receipt of this . . . at a price [of] £72 each all round at the sometime guarantee that they shall be works that Sydney may be proud of . . .
>
> Respecting the antique statues 7 ft . . . they will have to be modelled by a good sculptor from copies at the academy here and made by good men at about £200 at Sydney.[13]

Eight additional marbles were imported from Italy in 1883, including The Four Seasons, Rebecca at the Well (a popular statue of the period), La Ballerina, The Bather, and The Boy Extracting Thorn, as reported in the rather florid newspaper style of the day:

> During the last few days an additional attraction has been given to our Botanic Gardens, in the form of eight marble statues which now gleam white against masses of dark green foliage in various angles of the many walks which are so much appreciated by our citizens and visitors. They were purchased in Italy, and though not examples of the highest art — being indeed inexpensive works such as students and apprentices in Italian studios are mainly engaged in fashioning, their material gives them an immense superiority to plaster casts and, situated as they are, they have a very attractive appearance. . . All these statues are placed on well-finished freestone pedestals and Mr Charles Moore, the Director of the Gardens, has located them with excellent taste.[14]

It is believed that these statues were produced by, or under the direction of Charles Francis Summers (son of Charles Summers, sculptor), who was born in Melbourne in 1858 and went to Rome with his parents at the age of 11 staying there for 32 years. During this time he was friendly with two Popes, and given permission to take plaster casts of figures and carvings in the Vatican. From Rome he made several visits to Australia to exhibit and sell his marble reproductions of works of famous sculptors, including Canova, the greatest Italian sculptor in the early nineteenth century. The copies of The Boxers in the Gardens, of Venus in lower pond, and of Apollo Belvedere, all probably date from this period.[15]

In a letter to Sir Henry Parkes, 28 April 1881, Charles Francis Summers wrote:

> During our short interview in Melbourne I was not able to speak to you about some works that I heard would be required by your Government to be placed in the Public Gardens Sydney so I take the liberty of writing to you on the subject.
>
> From my position in Rome I am able to procure casts of all the most celebrated works in sculpture from the Vatican or Capitol or any of the galleries in Italy thus obtaining a better selection of choice works than can be procured by an Agent. I might mention that I obtained permission to mould one of the finest basreliefs in the Vatican and executed it in Marble for Sir George Verdon who is greatly pleased with the result of my labour. I have resided in Rome for the last ten years and have one of the largest Studios there which has been visited by most of your prominent Colonists many of whom can testify to my abilities and experience in works in marble.
>
> I hope to be in Sydney in a few days with the [earlier?] works you purchased from me and might I ask you not to conclude any arrangement with regard to the purchase or commissioning of works of Art in Italy till I have seen you for I am sure I can compete with anyone in the selection of suitable Figures, workmanship and fair charges.
>
> Thanking you for your Patronage.[16]

In a letter to the Gardens from a Miss T. Gidley in 1964 regarding the statues of the seasons, she states that her mother told her as a child they had been produced by a relative — a Miss Rosina Teresa McGuire, who also produced works in Tasmania.[17] Miss McGuire may well have worked in the studio of Charles F. Summers in Rome.

The indomitable J. H. Maiden (Director, Sydney Botanic Gardens, 1896-1924) had his say about statuary and fountains in public parks and gardens:

> We in Sydney have not been fortunate in our public statuary. Those of Queen Victoria, Prince Albert, Governor Bourke, and now of Burns, show that we have some worthy ones [and] that pretty little female figure by Birch surmounting the Levy fountain in the Botanic Gardens is far above the average of out-door statuary in New South Wales.
>
> The Governor Phillip memorial in the Garden Palace Grounds is huge and ponderous, and the tout ensemble silently suggests that we are an inartistic people . . . its artistic design is not quite satisfactory [and] the huge bronze recumbent figures are excrescences.''[18]

I cannot but agree. Maiden then went on to say that: 'Most of the statues of our public men are not works of art, and one sighs as one reflects on the large sums of money and sometimes ill-requited energies of public-spirited citizens which may have been expended upon them'.

Even in our own day, statues can still raise passions. In 1966 there was a question without notice in State parliament about the Phillip Fountain:

> Mr Jackett: Is the Minister aware that for many years this beautiful work of art . . . has long been among the great unwashed of this continent, with a pathetic trickle of water coming from its jets?
> Mr Chaffey: This is the sort of question that might well draw forth asomewhat facetious reply . . . I am sure (the honourable member) will find over the next few months that Governor Phillip's waterworks are rectified.[19]

Laughter was not recorded. The fountain was repaired by November.

In 1967 there were copious complaints about the 'modern' design of the fountain for crippled children that used to stand near the Palace Garden Gate.[20] And even more recently, in October 1989, a visitor to the Gardens (possibly an aspiring artist making an aesthetic statement!) pinned the following note to the Solander Bust: 'If this is an example . . . of the sculpture at the present time, then the standard has greatly deteriorated from what it used to be. I dare say Solander would be turning in his grave right now if he knew of this insult.'[21] Indeed the sculpture is somewhat gnomelike, but contemporary drawings and Wedgewood reliefs suggest that Solander was not what one may call a handsome man.

Maiden, true to his taxonomic nature went on to classify the statues:

> Statues may be grouped into three classes: those which embody the national or patriotic spirit, those which represent those somewhat vaguely known as 'public men', and the purely idealistic class.
>
> Amongst 'patriotic' statues I would enumerate those of Queen Victoria and Prince Albert.

Maiden then spoke of two proposed statues which never saw the light of day, and then returned to the 'public' statues:

> The statues of our public men include those of Governor Bourke, Lang, Dalley, Mort, Robertson.
>
> Idealistic statues are best represented by that of Birch's female figure already alluded to.

> Many of the statues in the Garden Palace Grounds and the Centennial Park are idealistic, as far as classification is concerned; but I hope that they will gradually be replaced by works of art, and then be broken up prior to some utilitarian application.[22]

And broken up they were, but not for such application.

The attrition of statues over the years has come from erosion and corrosion, from damage and vandalism, changes in public attitudes, plus changes in the Gardens planning and administration. Many have long since been removed, to languish in the 'graveyard' behind the Succulent Garden, others were returned to the Art Gallery of New South Wales, while still others have been dismembered or dispersed.

Reflecting on Albert's desire to uplift the 'lower orders', then reading the file that records the sorry litany of vandalism, makes me wonder sometimes why we even bother. And yet vandalism in the Gardens is quite mild when compared with that in Centennial Park (see Chapter 8). Sydney has always been a vital city with its downside ruffian streak, and as with graffiti there's a copycat element to vandalism.

Mr Kookaburra (having seen it all) told Snugglepot and Cuddle-pie that humans were 'as bad as bad'. The Blue-cap Wren confirmed the story as he had 'some relations living in the Sydney Domain' and he knew.[23] The descendants of that Blue Wren still live in the Gardens and Domain and would have seen a thing or two over the years.

Some writers have suggested that the 'knocking' element of the Australian character, in this case a 'knocking-off' or tall-statue syndrome, may come from the convict taint. Such views have been more fashionable of late, and may partly help explain our larrikin heart. To my mind, however, vandalism is more closely linked to the lack of availability of alternative outlets for the dissipation and sublimation of the social anger of young men in more conventional pursuits and — as more often than not it's the heads of the female statues that are carried off as trophies — as a misdirection of the male libido.

Statues in the Gardens have been in and out of vogue for various reasons over the years.

There was a 'morality purge' about 1915, as recorded in the *Sun* in 1930.

> 15 years ago representations were made to the Minister under whose control the Gardens were at that time, that a nude Venus, a nude Apollo, and other undraped statues, which were then in the Gardens, were a menace to public morals and should be moved from popular gaze. Orders were subsequently given that the statues should be taken from their places in the Gardens, and stored safely away where no sight of them could offend the eyes of the moralists of the day who chose to stroll that way. Records in the possession of the Department show that eight or nine years ago further instructions were given that the offending statues must be destroyed.[24]

A large number of statues were removed from the Gardens during the administration of Dr J. S. Beard in the early 1970s because of their deteriorated condition, and because they did not fit in with the ideas of landscaping at that time. By 1978:

> There was quite a gathering at the Botanic Gardens the other day (in the 'graveyard' behind the Succulent Garden) when these artistic figures got themselves into a huddle. Discobolus, The Bather, and a host of other distinguished statues looked as if they were holding a protest meeting against the vandals who forced them there.[25]

The statues in the 'graveyard' continued to create considerable interest over the years and the file is peppered with letters from various sources, requesting if they could purchase some of them. In 1979 a statue of Winter was made available on loan for use in the garden of the Chief Justice.[26]

The next Director, Dr L. A. S. Johnson did not 'wholly agree with the removals which were ordered by Dr Beard' and took steps to conserve existing statues''[27] The current Director, Professor Carrick Chambers, has established an active policy that existing statues be conserved and that selected statues and urns be gradually returned to the Gardens.

Given their historical associations it seems most appropriate that these Gardens retain some statuary, while eschewing clutter and not becoming a museum of the nineteenth century, for too many memorials would detract from the plantings, and make the Gardens look like Rookwood cemetery.

1

The Governor's Domain

Early Buildings and Significant Sites no Longer in the Botanic Gardens and Domain

Governor Arthur Phillip had been far-sighted enough to establish a 'Governor's Domain' by attempting to restrict to leasehold the alienation of land north of a line drawn on a map from the head of Woolloomooloo Bay to the head of Cockle Bay (Darling Harbour). Government House was originally to have been on the western side of Sydney Cove. The decision to build on the eastern side of the cove allowed Phillip to determine that the area east of the Tank Stream to Woolloomooloo Bay be exclusively reserved for the Governor's use. Phillip subsequently had a ditch dug and land between it and the water became known as the 'Phillip Domain'.

The Phillip Domain was originally reserved for exclusive use of the Governor. This was not always taken with good grace, but it did prevent early commercial development, although three windmills were established on the high ground of what is now the Palace Garden area. After the failure of the first farm the area was leased out as farmlets under 'permissive occupancy' to various people. Prior to that the land had been occupied by the Aborigines, who had hunted and fished, and gathered plant foods and shellfish in the area for thousands of years.

The Kangaroo and Dog Dance

Before the 'coming of the strangers' the Aborigines had camped in rock shelters in the region of the Botanic Gardens, and held their ceremonies at the head of Farm Cove and Woolloomooloo Bay. According to John Hunter, the natives of Sydney Cove seemed 'to have no fixed place of residence, but take their rest wherever night overtakes them: they generally shelter themselves in . . . cavities or hollows in rocks upon the sea shore, as may be capable of defending them from the rain'. Hunter then went on to say that, 'In the woods, where the country is not very rocky . . . a piece of the bark of a tree, bent in the middle . . . with a piece set up against that end on which the wind blows' was used as a shelter.[1]

From James Meehan's map (see p.35) we learn that the Aboriginal name for Farm Cove was 'Woccanmagully' and the tidal area along the harbour shoreline was the initiation ground of the 'Kangaroo and Dog Dance'.

On 2 February 1795, David Collins reported on an 'Extraordinary ''exhibition'' [at the] Head of Farm Cove where a space had been prepared by cleaning it of grass, stumps etc. [to produce] an oval shape 27 feet by 18, name ''Yoo-lahng''. The following exhibition was an initiation rite for young men/boys [which] involved the piercing of the nasal septum to hold an ornament, usually a bone or reed, the knocking out of a front tooth, and various dances including the ''kangaroo and dog dances'', the important part of the ritual being performed by the ''carradhis'' of the ''Cam-mer-ray-gal'' '.[2] During the dance a tribal elder with a grass effigy of a kangaroo on his shoulders was followed by another with a load of brushwood, signifying the animal's haunts. This was then deposited on at the foot of the initiates, who formed a circle, while the other warriors chanted and beat time on their shields. The dog dance was similar, with the actors crawling on all fours.[3]

The gunyahs have long since disappeared, as have the kangaroos, and dogs are no longer permitted in the Gardens. The kangaroo and dog dance is no longer performed on the tidal flats of Farm Cove, where until recently a ritual blessing of the fishing fleet was held each year.

Farmlets

Watkin Tench, writing in September 1789 of the problems of stealing from the gardens and stores, refers to 'the convict-huts on the eastern side of the stream, and the public farm', on the site of what is now the Middle Garden.[4] It is assumed from this that the Government Farm at Woccanmagully was still being used as a kitchen garden for vegetable cultivation. The convict huts would have been hardly more substantial than gunyahs — being probably no more than a slab-built or 'wattle and daub' room — with a dirt floor, and a bark (or thatch) roof.[5]

In spite of Phillp's orders that 'no ground within the boundary line [of the Domain] is ever granted or let on lease and all houses built within the boundary line are, and are to remain, the property of the Crown',[6] at least eight leases were granted between 1794 and 1806 for terms of 10 to 21 years.

We learn much about the developments in the Phillip Domain from Assistant Surveyor (of Lands) James Meehan's map, prepared for Governor Bligh on 31 October 1807. The position of Old Government House, with its garden and shrubbery, is shown. Over the area of what is now the Botanic Gardens is written 'Ground absolutely necessary for the use of Government House, but leases improperly granted on it, [it] is now improving'.[7]

In 1794 Phillip had given a 'permissive occupancy' of part of Farm Cove to Nicholas Devine, Superintendent of Convicts, consisting of two adjoining blocks (marked No. 95 on Meehan's plan) on both sides of the creek, extending to the mouth of the creek, in the region of what is now the Gardens Restaurant.[8] Two years later the Scottish martyr, Joseph Gerrald, took out a sub-lease on this farm, which contained 'a small house and garden, then to be sold'[9]. The position of this house and garden are not clearly defined, but they were probably in the region of what we now call the Upper Garden. Other farming leases in the region of the present Botanic Gardens and Mrs Macquaries Point are shown as follows:

No. 80, (higher up towards Government House) John Palmer, Commissary, showing a stone house and two windmills. Mr Palmer's garden is shown at 'Woolloomooloo House' at the head of Woolloomooloo Bay;
No. 92, Thomas Alford, Governor's Gardener;
No. 95, (two allotments already referred to) Nicholas Devine, Superintendent of Convicts;
No. 96, (two allotments) Nathaniel Franklin, Governor Hunter's butler, (who suicided 1798)[10];
No. 97, Frederick Markett;
No. 98, (two allotments) James Callum (Callund or Callan);
No. 99, (two allotments) Alexander McDonald.

The merchant Michael Hogan was granted a lease in 1796 in the Tarpeian Precinct of the Domain by Governor Hunter for 2*s*. 6*d*. per year. His lease was for 14 years for the land from the back of the fence of (new) Government House down to the water at Sydney Cove. It was terminated by Governor Bligh and there is no record of any buildings in the area at that time.[11]

Farming was not the only occupation carried out in the region, for in 1805 one Michael Hayes, of Farm Cove, was convicted of operating a 40-gallon still.[12] Hayes' neighbours at the time were John Cummings and John Archer.[13]

John Anson later had a lease on the eastern point of Farm Cove; thus 'Ansons Point' (Mrs Macquaries Point) was advertised as being for sale on 29 May 1808:

> To be let or sold by Private Contract, a truly valuable leasehold farm beautifully situate at the Eastern Point of Farm Cove, the property of and formerly in the occupation of John Anson; the whole containing 11 acres and upwards; with excellent pasturage and every accommodation for a stock farm; abundantly stocked with various choice fruit trees, and in all respects worthy of attention. For particulars, apply to J. Anson, Carpenter, Back Row East.[14]

Windmills and Bakery

There were three windmills in the Domain in the early days of the settlement, on the high ground in the region

Boston's Mill in the Domain, *RBG Library*.

of what is now the Palace Garden grounds. They were Boston's Mill (current site of Huntsman and Dogs statue), Palmer's Mills and Bakehouse (site of Conservatorium of Music), and Kable's Mill (built by Nathanial Lucas at the site of Shakespeare Monument near the State Library of New South Wales).[15]

John Boston, surgeon and apothecary, who arrived in Sydney as a settler in 1795, had his mill operating by the late 1790s. The mill equipment was removed from Boston's Mill in 1828, and it was turned into a storehouse, to be demolished about 1840. John Palmer had arrived on the *Sirius* in 1788. He became Government Commissionary in 1791 and built a large house in Woolloomooloo (near where Palmer Street runs today). Although Governor Phillip had said that there were to be no buildings in the Domain, Governor King had given a lease to John Palmer and a Mr Riley in 1802, to help them carry out what was essential work for the colony.[16] They constructed their mill in that year, with a bakehouse immediately to the north. Between them, on little more than three acres of land, they ultimately ran two windmills and a bakehouse.[17] In 1804 Nathanial Lucas was granted a lease of 14 years and built his mill, which was leased to Kable and Underwood until 1811, when it was advertised for sale. It was again let in 1816, and moved to a new location in Darlinghurst in 1820.[18]

When Governor Macquarie arrived at the end of 1809 he was less than enthusiastic at having such operations 'only 160 paces from Government House' which prevented the construction of his wall around the Domain. Macquarie, becoming obsessive about the privacy of his Domain, finally took advantage of the expiration of Palmer's lease (which had been mortgaged to Robert Campbell who assigned it to Messrs Fairlie and Company and Messrs Ferguson and Company of Calcutta in 1813) to pull down the mill and bakehouse.

Riley was in the bakehouse, removing a batch of bread from the oven, when the Secretary, John Thomas Campbell, flanked by two guards, brought the news in November 1813. Palmer was called from the neighbouring windmill to hear that Macquarie had given them just 16 months to clear their debts and move. The land was resumed, the buildings were demolished in 1815, and Palmer became insolvent.

The foundation stone of the new Government stables (later to become the Conservatorium of Music) was laid on the site of the bakehouse in December 1817. Twenty-four years later the court was still considering a claim by Palmer and Riley, even though John Palmer had died six years earlier.[19]

The last located reference to a windmill on the Garden Palace Grounds is in a letter from John McLean, Assistant Superintendent of the Gardens, to the Colonial Secretary on 14 February 1834, in which he complains:

> I beg leave to state for the information of His Excellency the Governor, that the newly made plantation leading to the Domain is entirely destroyed with horses. There is a man of the name of Still who has taken possession of the Mill [probably Kable's Mill] and the grounds attached to it. He keeps a watchman, and when I send men to clean the grounds he threatens me to put them in the watchhouse. The watchman in charge tells me he will send as many horses as he thinks proper to destroy all the young trees. I beg leave to know how to act.

The Old Well near Conservatorium of Music Entrance, 1917, *RBG Library*.

Lime Kilns and Wells

Composite maps of early Sydney have indicated the presence of lime kilns in the region of what is now Bed 130, near the Opera House Gate.[20] Lime pits or kilns (like the one still intact on Goat Island near Andersons 'Chair' or 'Cave') were used to roast shells to make lime for mortar production for buildings and wall construction. A photograph in the Gardens Library also identifies an ancient well in the region of the Conservatorium of Music. It is assumed there would have been numerous other wells associated with the buildings in the area.

Early Houses

The first Government House in the Phillip Domain was a prefabricated canvas (oilcloth) tent that had been brought out with the First Fleet.

Old Government House, a two-storey Georgian building, designed by James Bloodsworth, a convict master bricklayer, at the corner of Phillip and Bridge Streets, was set in a garden and orchard that extended down to Sydney Cove. Its foundation stone was laid on 15 May 1788, and Governor Phillip moved in during April 1789. The Aborigine Arabanoo, who had been adopted by Phillip, died there of smallpox on 18 May 1789 and was buried in the grounds.

The Bridge Street building (in addition to the house in Parramatta Park) was the vice-regal residence for more than 50 years. It was added to over the years, and considerably altered, while steadily worsening in condition, and finally demolished in 1845–6, at the time of the move to the new house. The engraved copper foundation plate of Old Government House was found in 1899 by workmen digging a trench in Bridge Street.[21] Its foundations were rediscovered in 1983.[22]

In 1791 Governor Phillip had a substantial brick hut, some 12 feet square, built for Bennelong (a member of the Guringai tribe) at Bennelong Point, somewhere in the vicinity of where the Opera House now stands. First named Cattle Point as the stock from the First Fleet had been landed there, it was a rocky peninsula with a small island at its tip. The cattle were later moved to Farm Cove.

Billy Blue, a part-Jamaican emancipist who oversaw the government salt-house (producing salt from sea water) also lived on the eastern side of Sydney Cove at Bennelong Point. The salt-house was not very efficient, producing only three to four bushels of salt in more than as many weeks.

By 1807 a number of houses had encroached on the Phillip Domain. The houses of Nicholas Devine and Joseph Gerrald (the same house, or possibly one of the earlier convict-huts?) have already been mentioned. Governor Bligh, interpreting Phillip's earlier edict literally, decided to remove them, and issued the following general order on the subject on 23 July 1807:

> Whereas His Excellency Governor Phillip, did, by instructions from His Majesty's Ministers, draw two lines of demarcation in the vicinity of Sydney, within which no leases or grants of land for building were to be given, the said land being the property of the Crown.
>
> And whereas a number of houses adjacent to Government House, to its great annoyance, now occupied by David Dickinson Mann, Abraham Ramsden, John Apsey, William Kimber, John Shea, Ferdinand Meurant, and others, within the said limits, have been built on land particularly marked out as making part of the Domain of the Governor's residence; the Governor is pleased to extend the notice already given to the first day of November next; at which time they, the present occupiers, are required and directed to quit possession of the said houses, taking away or disposing of their materials, the said ground being wanted for Government purposes.
>
> His Excellency the Governor-General is ready, on application to that effect, to grant permission to the said persons to build on such other ground, unoccupied, in the Town of Sydney as may not interfere with his arrangements on that head.[23]

It is assumed that all the other houses in the Domain were removed at about the same time, which may explain why John Anson was so anxious to sell in 1808.

D. D. (David Dickinson) Mann's house is shown on Meehan's map (1807),[24] at the rear of the Old Government House and slightly to the north of where Bligh Street now runs into Bent Street. Behind that were the houses of the Judge Advocate, the Chaplain, and the Surveyor.

It is probable that some, if not all, of the other houses mentioned in Bligh's edict were contiguous, as a row of houses ran in a north-east and south-west direction, behind old Government House, at the top of Bridge Street.[25]

Bligh's attempt to reinforce Phillip's edict by redefining the 'ditch' and by progressively removing leasehold farmers from the Governor's Domain created quite a stir, as recorded by Surgeon Harris, writing to Mrs King, wife of the late governor, on 25 October 1807:

> He [Governor Bligh] has so changed everything about Government House, that you would be entirely lost in it, but I cannot point them out to you — the vast alteration. The shrubbery has also undergone a thorough change — no grass now growing in it, all laid out in walks with clumps of trees. Even the poor tomb of young Kent is annihilated. All the rocks in the garden are blown up and carried away. Not less than 80 or 90 men have been constantly employed since you went away, for these purposes. Carriage roads are now all around Bennelong's Point, and down about Farm Cove, all ditch'd in and no thoroughfare allowed. All dogs are ordered to be shot at, at which his secretary and bodyguard have had much amusement. He has ordered Murant's [Meurant's] house and all that row to be pulled down, and which has been done to the total ruin of those poor wretches. Even the house of D. D. Mann has been ordered down.[26]

Bligh had made many enemies and was even attacked one night in his own garden by his head gardener's dog. The head gardener, 'a huge violent man [called] William McLaglen', apparently mistaking Bligh for an intruder, weighed in in the dark as well. The next day Bligh penned the order stating that 'All dogs found in the garden henceforth will be shot'.[27]

Evidently Bligh's men made a clean sweep of the Domain, including the grave of Arabanoo and other tombstones (which may account for the loss of Gerrald's stone see p.127) and the 'tomb of young Kent'.[28] The tomb referred to was probably that of John Hunter Kent, born 8 November 1795, died 18 May 1796, son of Lieutenant William Kent, master of HMS *Supply*, which was probably buried along with all the other 'encroachments'. In 1847, while workmen were preparing the carriageway from the eastern end of Bridge Street to the gateway leading to the western entrance of the grounds around New Government House, they discovered the tombstone of John Hunter Kent, near the site of Old Government House.

The first place of public worship in the Colony was adjacent to John Kent's tomb, where a 'rude wooden building of a temporary description' had been built.[29]

On 1 August 1812 the houses of Nicholas Devine, in the Government Farm area near the Spring and Tanks in the town, were surrendered to the government.[30] This would tend to indicate that the original house of Nicholas Devine (and Joseph Gerrald) was in the region of the headwaters of Botanic Gardens Creek.

Part of old Macquarie Wall at Woolloomooloo, *RBG Library*.

Walls and Fences

It has been noted earlier that Macquarie was not happy to have windmills operating so close to Government House. When he became governor in 1810 his wife had objected to being stared at in her garden — an area of shrubbery laid out with walks and trees by Governor Bligh. She liked to walk in the garden but had felt herself 'overlooked' by the mills and the bakehouse, the mills having been built on the higher ground.[31]

Macquarie's attractive young second wife, Elizabeth, had been sickly almost from the day she had arrived in the Colony and may well have been suffering from morning sickness for much of the time as she had become pregnant in June. The Macquaries' second child, also called Lachlan, was born on 28 March 1814.[32]

At any rate, Macquarie enthusiastically set about containing the Governor's 'Demesne' (the old spelling of Domain officially in use until about 1816) with a series of sandstone walls and paling fences, motivated as much by a desire for privacy as the other governors had been in what was still a penal colony.[33]

Work began on the gigantic wall in August 1810 under the supervision of the old stonemason, Thomas Boulton. It was of 'good tough square-hewn stone on one side, and good rough work on the other', according to the agreement, and grew to three metres in height, topped with wrought ironwork. Crowds gathered and asked Boulton questions but he gave no answers. Macquarie came out each day to inspect the progress, tapping his military cane in his hand behind his back. After four months the wall was 290 metres long and Boulton was given an interim payment in rum.[34]

On 6 October 1810 Governor Macquarie had set aside a portion of the old Phillip Domain for Hyde Park and for a General 'Rum Hospital'.[35] His 'Rum Wall' was never to engender the same affection as the hospital. As of 1810 the Domain, although established as an exclusive reserve for government use, had not excluded the public.[36] People resented the intrusion on what they regarded as public land and viewed the construction with the same repugnance that later generations were to have for the Berlin Wall.[37]

In October 1811 John Mould was contracted to make posts and railings for the dwarf wall round a portion of the Government Domain as follows:

> Sydney, 28th October 1811. Jno. Mould, carpenter, doth agree to put up and complete eight hundred and eighty seven feet of pallisading [*sic*] on the dwarf wall, for Government House, to the end of Bent Street, and agrees to complete the whole in a workmanlike manner, for the sum Sixty Pounds sterling.[38]

R. H. Cambage thought that this was in the area of Old Government House and this has since been supported by J. F. Campbell, who said that the 'fencing . . . along Bent Street is described as palisading on a dwarf wall'.[39] In his evidence to a 'Report of the Parliamentary Standing Committee on Public Works Centennial Park Reservoir' in 1892, Charles Moore, in responding to Question 381, refers to a dwarf wall running in front of Richmond Terrace.[40]

By 1812 the Domain had been entirely enclosed from the head of Woolloomooloo Bay to Sydney Cove with stone walls and paling fences, with the exception of the parts leased to Riley and Palmer. This did not prevent people climbing over the wall and breaking it down in places — to cut down and burn the shrubbery, destroy young trees, steal plants, loam and stone — and take home wooden palings for their fires.[41]

On 17 October 1912 a government public notice was issued by Macquarie, prohibiting the grazing of animals, quarrying of stone etc. in the Domain:

> The whole of the Government Domain at Sydney extending from Sydney Cove to Woolloomooloo Bay, being now completely enclosed by stone walls and palings, except that part which is at present under lease to Mr Palmer and Mr Riley, where their windmills and bakery are erected, notice is hereby given that no cattle of any description whatsoever are, the present date, to be permitted to graze or feed on the said Domain, those belonging to Government only excepted; and any horses, sheep, asses, pigs or goats which may after this notification be found trespassing thereon will be taken up and impounded for damages.
>
> The public are further to take notice that for the future no stones are to be quarried, or loam dug, within the said Domain, for other than Government purposes, unless by special permission obtained from His Excellency the Governor; and none of the wood or shrubs growing within the said Domain are on any account whatever to be cut down or otherwise destroyed, on pain of prosecution for felony, and no boat, except those belonging to the Government, is to land in Farm Cove, or any other part of the shore bounding the Domain, except at Bennelong's Point, on pain of being forfeited, and the persons who have so brought them there severely punished for trespassing on the Government Domain after being thus cautioned by public notice.[42]

Since the proclamation apparently did not prevent people climbing the wall at night to hunt or collect wood, the governor issued repeated orders, but to no avail. Harsher and harsher punishments were handed out, but people still climbed the wall.

Macquarie finally decided to put a stop to this trespass by flogging people who climbed the wall without reference to a magistrate. In 1817 John Hersall, Daniel Reid, and William Blake (not the poet!) received 25 lashes each for trespassing on government ground.[43] However, Judge Bent complained to Earl Bathurst that the Governor was taking justice into his own hands, and Macquarie was forced to bow to public opinion.[44]

In a notice of 6 July 1816, referring to trespassing and the destruction of property in the Domain, the following passage occurs:

> These orders, however, are not meant to extend to prohibiting the respectable class of inhabitants from resorting to the Government Domain as heretofore for innocent recreation during the day time; the road some time since constructed round Bennelong's Point furnishing easy access to that quarter, and the gate and the stile at the East end of Bent Street offering free admission in that direction.[45]

Sydney's Botanic Gardens had been born, although they were not officially proclaimed.

After the death of Allan Cunningham in 1839 the carriage road north of the Macquarie Wall, still to be found in the Gardens, was fenced with palings, thus shutting out the view of the harbour. This was done to enable the Lower Garden, laid out in the early 1830s, to prosper. Soon after, Mrs Macquaries Road was diverted and the fencing of the Lower Garden completed.[46]

By 1867 the south-eastern part of the Domain had been altered almost beyond recognition — the 'eastern boundary now [being] marked by a neatly-built dwarf parapet wall, with iron palisades'.[47]

In 1905 80 feet of the Macquarie Wall behind Sydney Hospital was removed and 'the stones were partly used in the building of a picnic fireplace in the Domain, and partly in facing a portion of the sides of the upper reaches of the creek in the Botanic Gardens'.[48]

Nine years later Mr S. E. Lees, reflecting on the changes in the Domain, stated:

> The contour of the grounds has not altered very materially. Still there have been changes. For instance, the high wall that skirts the western boundary at the rear of Parliament House and Sydney Hospital used to extend in a fairly straight line as far as Palmer Street. The only interruption in it was for the entrance gates near St Mary's Cathedral and the lodge adjoining. The wall from the lodge to the Palmer Street gate cut off a considerable part of the Domain fronting St Mary's Road, which was enclosed by a picket fence. An entrance to this part stood at the corner of St Mary's Road and its junction with the main road leading to the present gates alongside the Registrar-General's new offices.

Lees then went on to say that:

> A few yards from the wall in this part [of the Domain] was an old underground brick drain [presumably the water link from Hyde Park to the Gardens — see Robert Dawson Memorial, p.123]. In several places it had subsided, and the covering earth had fallen in with it. The crown of the oval-shaped brickwork had broken in, leaving an unsightly receptacle for rubbish. [By 1914] that portion of the wall had long since been removed, the drain filled in, and the ground thrown into the Domain proper.[49]

In 1914 the dwarf walls and railings were still to be found on the Woolloomooloo side of the Domain.[50]

As of 1988 the only remaining part of the (renovated) Macquarie Wall system is the wall in the Gardens between the Restaurant and Lion Gate Lodge (see Chapter 4).

Buildings after 1816

The traditional foundation date of the Gardens is taken as 13 June 1816, the date of completion of Mrs Macquaries Road. A number of buildings that no longer exist in the Gardens and Domain were constructed after this time.

Between 1817 and 1819 Governor Macquarie replaced an earlier redoubt built in 1798 on Bennelong Point with a stone fort (Fort Macquarie), mounting 15 guns. Fort Macquarie on Bennelong Point remained manned until 1902, when it was demolished to make way for a tram shed. This in turn was demolished in 1959, when trams were being replaced by buses.

Prior to that a small landing place had been built there for his private use — now known as 'Man-o'-War' steps — on what was then known as Port Lachlan.[51]

From 1856 to 1870 there was also a fort on Mrs Macquaries Point (Macquarie Point Battery — not to be confused with Fort Macquarie) immediately behind Mrs Macquaries Chair.[52]

Cartwright's plan of 1816 shows four buildings in the region of the Botanic Garden, one with two outhouses on the eastern side of the path, and one on the opposite side. One of these (in the region of Lawn 2), a little north of where the Levy Fountain now stands, was Mr Fraser's home. Governor Macquarie claimed credit for the construction of this cottage, which therefore must have been built between 1810 and 1816: 'a Government Garden made on Farm Cove, consisting of five acres of Ground, enclosed partly by a High Paling, with a Brick House for the accommodation of the Chief Gardener'.[53]

A letter from Allan Cunningham to R. W. Hay in 1832 would indicate that the original structure was quite modest:

> As the house of the late Superintendent [which was originally, most probably simply intended as a vegetable or seed room, and not to be a dwelling house] is small and internally very inconvenient, it is respectfully hoped that it may be enlarged, by an addition of wings, and otherwise subdivided, so as to render it habitable and comfortable to his successor.[54]

The key to the 1829 plan (Burford's map key refers to 'Botanic Gardens' and 'residence of Mr Fraser'), and the 1833 and 1867 plans show the sites of official residences. The 1833 plan shows a building of three sides (with an inside quadrangle) on the site of the old Museum (now Visitor Centre).[55]

In the mid-1980s, excavations just north of the gate in the Macquarie Wall unearthed an old convict drain that appeared to be coming from the direction of the Visitor Centre. The drain had a hemispherical cross-section carved into the solid sandstone, with a flange on both sides, along which ran a double row of convict bricks, with shell-lime mortar. Capping stones of solid sandstone rested on the bricks.

Richard Cunningham, when he arrived in January 1833, took lodgings in Sydney to allow suitable renovations on Mr Fraser's house to be undertaken. Further renovations then occurred in 1837, before Allan Cunningham occupied the house.[56] On 1 July 1833 a minute from Richard Cunningham requested: That the Colonial Architect be instructed to furnish an estimate of the expense of a small range of offices for the use of the Department; of a stable, a store room, and a cartshed, by which I beg to observe much expense would be spared to the Department under the head of repairs.[57]

In 1837, a seed room and barracks for lodging the men in the Botanic Gardens was erected:

> It stood on the site of what is now called The Anderson Building, and is indicated on the 1867 plan of the Gardens. The original purpose of the barracks was to provide accommodation for young convicts so that they could be trained in the 'art of gardening' . . . [and] . . . obtain good employment when they obtained their Ticket of Leave.[58]

In 1849/50 Richmond Villa was built in the Domain by Mortimer Lewis, Colonial Architect, as his private residence.[59] Previously Richmond Terrace (built 1814) and Domain Terrace had been built on part of the old Phillip Domain.

A description of the buildings from this area from the 1850s by S. E. Lees identifies them both:

> A little south of the Macquarie-street gates, and built on the vacant land now forming the street frontage to the Mitchell Library, stood the old barrack-like building [previously the light-horse barracks], which abutted right onto the street alignment. This was known as the School of Industry for Girls. Behind this school, and at the eastern end of the lane that flanked its southern side, stood a row of small tenements known as Domain Terrace. On the north-west corner of this lane there stood an old cottage abutting far out from the building line. This had a garden enclosed by a picket fence. The Parliament House bowling green now stands there. All these have gone, but the buildings fronting the Domain, and known as Richmond Terrace, still stand.[60]

Richmond Terrace was demolished in 1935 to make way for the expansion of the State Library of New South Wales. Richmond Villa, to the south of Richmond Terrace, survived this demolition, possibly because it was being used by the government at the time.

Mortimer Lewis never lived in Richmond Villa. He was declared insolvent on 12 November 1849 and the property went to George John Rogers, a Sydney solicitor, who sold it to Samuel Peck.[61] Subsequently numerous people owned the property. By 1893 it was being used as quarters for the Librarian, and as a storage for parliamentary records, magazines and bound newspapers. It is believed that 'The True Blues' — forerunners to the Country (National) Party — occupied the house from as early as 1922; by 1932 they were using the ground floor room as offices.[62]

Plans for the new Parliament House finally encroached on the site and Richmond Villa was moved to Kent Street in 1975, to become the headquarters of the Society of Australian Genealogists.

The land on which Domain Terrace was built had been granted to J. Buckley. This terrace was demolished in the 1920s and the eastern part of the Parliamentary Bowling Green partly occupied this site.[63]

Also in 1849:

> The barrack, which was formerly occupied by the prisoners, has also been put in a tolerable state of repair, and divided into five apartments, by means of brick walls, which are now used as store and lumber rooms, and partly by the workman for domestic purposes.[64]

The buildings were certainly not elaborate in those days, for in the Report of the Select Committee to the Botanic Gardens, 1855, the old lecture room was described as follows: 'By taking the bricks out of an old drain, I put up two walls to the old barrack; and then I got a sum of money from the Colonial Architect to roof and plaster the place'. In 1870 Charles Moore reported that:

> The office or library, the quarters of the Bailiff and Propagator, adjoining the store rooms, the Overseer's house, and even that in which I reside, are all unsightly. With the exception of the latter, all are old, badly built, in wretched condition, damp, most unhealthy, and quite unworthy of the place. To level every one of them with the ground and replace them by others more suitable, in better positions, and lay out the ground which they now occupy in an ornamental manner, are works which urgently require to be done.[65]

In 1878 a general clearance of all these buildings was made. The stables for the horses employed in the Gardens and Domain were transferred to the old Kitchen Garden, out of sight and away from public resort.[66] The old barracks,

> in which were the office, herbarium, library and quarters

Director's House, Botanic Gardens, 1855, showing Charles Moore in foreground and his wife Elizabeth and his niece Isabel, in light dress. Demolished in 1875 it was situated between the site of the Levy Fountain and the Palm House (Glasshouse No. 10), *RBG Library*.

> for employees, as well as the stables attached, were pulled down, and new buildings of a substantial character erected instead. These new buildings consist of a museum, lecture room, library, office, storeroom and two separate dwellings for men of the establishment.[66]

A photo of Fraser's enlarged home shows Charles Moore standing rather stiffly (to obtain the necessary exposure) in his garden, with his wife Elizabeth and niece Isabel on the balcony. This is the same house in which Allan Cunningham, James Anderson (Assistant Superintendent May 1838–April 1842) and William Robertson (Superintendent May 1842–July 1844) all died.[68]

The new Director's Residence was constructed in 1874, and the old residence demolished in 1875. The overseer's residence was demolished in 1878 and replaced with a new building on another site.[69] An article in the *Town and Country Journal* on 6 April 1878 refers to the overseer's house: 'On the left of the path [in Upper Garden] heading north [along the main pathway] is the temporary office, formerly the head gardener's house; and this, too, is soon to disappear beneath the hand of improvement'.[70] This same article referred to stonework being prepared for the new Museum, Library and Lecture Rooms.

According to Walter Froggatt, another of the old gardener's cottages in the region of the Governor's Bathing House was demolished in 1912. As of the time of writing (1931) the roughly carved date '1830' could be seen on an adjacent rock.[71] In 1988 I was unable to confirm this date, as all possible rock carvings were badly eroded. The Annual Report of 1913 confirms that: 'The cottage [lodge] on the western side of the Gardens, formerly occupied by the Head Gardener, Federal Government House, was demolished in March [1912] because its occupation of an exceptional site blocked necessary landscape improvements. It was erected in the year 1856, and was of brick'.

More recently, the Government House caretaker's cottage overlooking the Opera House from on top of the Tarpeian Way, was demolished in the early 1980s. The proposed landscaping plan for this site was reported in the 1981–82 Annual Report.[72]

A modest house-like Gatehouse at the Palmer Street entrance, south-east of Domain (St Marys Lodge), at the southern end of what is now the Domain Parking Station, has also been identified.[73] This should not be confused with Domain (St Marys) Lodge (on Art Gallery Road — see Chapter 4), built in 1835 and still in use.

Another old cottage (gatehouse), the former residence of one of the 'Domain bailiffs', was pulled down prior to 1842, when the road was widened where Governor Bourke's statue was to be placed (where the Shakespeare Group now stands).[74] The site established for the statue was 'just inside the paling fence opposite to where the cottage formerly stood'.[75]

Mrs Charles Meredith, writing of the Domain in 1839, states: 'Lady [*sic*] Macquarie had this Domain laid out after her own plans . . . and lodges built at the entrances'.[76] Apart from Domain (St Marys) Lodge it is not known what other lodge(s) she may have been referring to, but lodges would have been required after 1831, when the roads in the Domain were opened to carriages.

The old and unsightly houses of Messrs Farmer and Richard Hellings (Corporation Baths), and fences associated with the old Woolloomooloo Baths were demolished in 1902, and the area improved by landscaping.[77]

Early Glasshouses

According to Walter Froggatt:

> In 1808 there was a glasshouse known as the 'Pinery' in the Botanic Gardens, in that part known as Old Government House Garden, for on 9 August of that year Governor Bligh instructed Thomas Alford, the Government Gardener, to represent the want of glass for pineapple plants. There is a record on 28 November, 1829, of a terrific hailstorm, which broke all the glass in it. It was probably situated opposite the present Botanic Gardens office building. It was enlarged in 1830, repaired in 1833, and on 1 December, 1844, it was demolished.[78]

In 1822 the Director, Charles Fraser, unsuccessfully sought funds for a conservatory, 'length, fifty feet, greatest height, 20 feet, greatest breadth, 20 feet', so that he could cultivate 'many Tropical Fruits, culinary, medicinal, and other Plants, so as to perfect their Seeds'.[79] However, by 1846:

> In what was formerly merely a frame yard, beside the entrance to the upper Garden, there is now nearly completed, a hot-house.
>
> This was much wanted for exotics, which could not be preserved, whatever care was bestowed, on their arrival here, from want of such a place. It is feared that many of the seeds which Dr Leichhardt brought with him will be lost. The plant shows itself, but only to wither again, not being sufficiently protected . . . so also, valuable collections sent from India are in risk of being lost. A spaceous greenhouse on the opposite side, behind the [Whale's head] 'Rockery', would be very useful, as well as ornamental addition to this part of the gardens.[80]

Three years later, in 1849, a large plant house was built in the region of the old brick pit (location unknown, although it may have been in the vicinity of the lime kilns), some 38 × 12 feet and 8 feet in height. It was covered with a span roof with glazed sashes, and was used for propagation of plants and raising seedlings.[81]

By 1854 more 'neat glass-roofed brick pits, suitable for the cultivation of small and tender plants' were constructed,[82] and 'In 1871 Thomas Mort presented a valuable collection of hothouse plants to the Gardens, together with a greenhouse. This was rebuilt on a site in line with the Botanic Museum, a little to the west.'[83] This glasshouse, in the form of an 'irregular octagon', was fitted up by William Huntley, who worked for 26½ days at 8 shillings a day. This new building was 'a conspicuous ornament near the principal entrance' and by the Spring, was 'daily accessible to the public'. Plants initially provided by the donor included ferns, begonia, azalea, rhododendron, hibiscus, anthurium, dracaena and coleus.[84]

Important works were carried out in 1899 in connection with the plant-houses, which

> were all in a serious state of disrepair; in fact, two, if not three, of them were in such a condition that I was afraid I should have to exclude the public . . . The old houses have been practically rebuilt with the exception of the brickwork. The whole of the houses have been repainted, and now look bright and clean.[88]

A 'clean sweep was made of the dilapidated structures' (old glasshouses, propagation and plant houses) on the western side of the Herbarium in 1907, as indicated in Maiden's 1903 map, to make way for a pergola and Rose

CSIRO Glasshouse on site of Robert Brown Building, *c.* 1936. The Cunningham Building (Director's Residence) is in the background, *RBG Library*.

Garden. There were 'five very old glasshouses [in this region], with potting sheds and boiler houses, the chimney of which had for years been in danger in regard to fire and smoke so near the Herbarium'.[89]

Three new 'plant houses' were erected by the Government Architect's Department in 1909, as well as a spacious boiler-house with improved hot-water service.[90] The remaining old glasshouses were all demolished in 1915.[91]

In 1936 a research glasshouse was built for the CSIRO on the site of the Robert Brown Building.[92] It was subsequently surrounded by six temporary buildings, all of which were used by the CSIRO. One of these, the seed room, was a brick and concrete bunker, built as an air-raid shelter *c.* 1940. It was used for some time by CSIRO as an annex, where they conducted experiments on the biological control of fruit fly.

The building, under a fine fruiting specimen of *Macadamia tetraphylla*, fronted the street and the old car park on the site of the Robert Brown Building, and looked towards the Cunningham Memorial stone and the old Preparation Room (site of palm bed between Brown and Anderson buildings). The old seed room and other buildings were all demolished in the late 1970s, when the new Herbarium was being built.[93]

Rustic Bridges, Pavilions and Bandstands

Rustic bridges and pavilions were a popular feature of Victorian gardens.

Old photographs show the curved bridges over Botanic Gardens Creek. In Patrick White's *Voss*, set in the late 1840s, the explorer (based on Ludwig Leichhardt) had discussions with Mr Palfreyman (to my mind a character based on Allan Cunningham — whose brother Richard suffered a similar fate to that of Palfreyman) on 'an ugly ornamental [rustic] bridge looking down at the dead water-lily leaves'.[94]

Two octagonal pavilions were built in the Lower Garden in 1863, of ironbark saplings with roofs of thatched grass tree leaves in accord with the style promoted by J. C. (John) Loudon and his wife Jane. The Loudons produced the first accessible gardening books — the progenitor of contemporary garden books and magazines — at a time when bright coloured flowers were coming into England, and were the 'high priests' of the high Victorian style of carpet bedding, rustic benches and pavilions.

With the founding of *The Gardener's Magazine* in 1826, John Loudon became the first horticultural journalist. In 1829 he inaugurated a *Magazine of Natural*

Rustic Bridge over Botanic Gardens Creek, 1882, *RBG Library*.

History, but he is best remembered for his books, especially *The Encyclopaedia of Gardening* (1822) and the monumental *Aboretum et Fruticetum Britannicum* (1838).[95] Jane Loudon published *Gardening for Ladies* (1840) and *The Lady's Country Companion* (1845).[96]

The western pavilion was in the northern end of what is now Bed 85, the eastern one in Lawn 36. In 1902 the 'much decayed' western pavilion was demolished. This may have been replaced, as both pavilions are shown in the 1912 map of the Gardens (but interestingly not the 1907 vista pavilion in Lawn 54). The thatch roof of the eastern pavilion was replaced with redwood shingles in 1906.[97] The western pavilion had been removed by the 1914 map. The eastern pavilion was repaired again in 1915, having been ravaged by termites.[98] It is shown in the 1921 map and was demolished between 1927 and 1954.

In 1858 a bandstand was erected in the Lower Garden on the large lawn (Lawn 24), which is still known as the Band Lawn. The splendid music of the 11th Regiment used to play there on certain afternoons and was a great attraction. This band was succeeded by that of the 33rd Regiment, which in turn was followed by the band of the 77th Regiment.[99] In 1866 the bandstand was roofed with iron. It was demolished in 1911.[100]

There was also a pavilion in the Domain in 1878. Mr G. L. Carter, lessee, entered into an arrangement with Professor Rufus Gibbon Wells in which they would share the net profits from 'captive' balloon ascents in the Domain.

At 10 p.m. on a Saturday evening in January 1878, a young sailor James M. Stewart volunteered to go up in the balloon, along with a would-be aeronaut named Francis, who jumped out of the basket as the balloon was leaving the ground.[101] Unfortunately another young man, Charles Tapley Harden, who worked at the Government Printing Office and had been holding on to the side of the basket, was carried into the air and fell. He subsequently died from spinal injuries. At the same time one of the ropes struck the Professor across the face, knocking him out and blackening both his eyes.

As the balloon rose rapidly, the tethering rope snapped and Stewart was tipped from the basket. His experience as a sailor probably saved his life. He managed to grasp the trailing rope as he fell, and climbed back into the basket — some 1,000 feet above the ground — losing his hat and gold watch in the process.

After that it was almost all plain sailing. In an age before aeroplanes, Stewart marvelled at the expanding city views, and especially at the lines of gas lamps in the streets below. The balloon headed in the direction of Governor Bourke's statue (current site of the Shakespeare Memorial) when the wind changed, taking it over Mrs Macquaries Chair and Fort Denison to the North Shore, where the wind changed yet again, bringing it back in an almost perfect circle.

When the air cooled and the balloon began to descend, it become entangled in one of the trees in Government House grounds, and upturned the basket again. Stewart escaped with only a few scratches to his face by climbing 20 feet down one of the guy ropes, to the amusement of the guests who happened to be attending a garden party at that time.

Eventually the balloon was secured, with the assistance of some sailors and Government House staff, and Stewart went 'down the town' with a number of friends.

A bandstand also once stood on the site of the Restaurant on the Park in the Phillip Precinct of the Domain. It was built by the Department of Works in 1897[102] and was the scene of many concerts.

> On Monday afternoon, too, between the hours of four and five, the Domain is resorted to by the fashionables and

> unfashionables of Sydney, eager to listen to the harmonious strains of the band of the 99th Regiment. There you may see the elegantly-dressed ladies, in their splendid carriages, on the doors of which some favoured gentlemen are wont to lean, and chat with the admired ones within. There, too, are settlers from the interior, whose immense beards and moustaches, monkey-jackets, and cabbage-tree hats, give them a Robinson Crusoe-like appearance . . . It is our lot now to see him mounted on one of the colonial much-labour-enduring breed of horses, or driving tandem round the Domain. There are not a few gentlemen on horseback or on foot, while crowds of the humbler classes are resting on the greensward, listening with pleasure to the lively polkas, and enjoying the charms of the beautiful scenery [until] we hear the thrilling swell of the National Anthem.[103]

In 1855 the military band (11th Regiment) used to play from 4 to 6 p.m. every Tuesday afternoon, through the kindness of Colonel Bloomfield, on the grassy space of the Domain

> in front of a fine row of houses [Domain Terrace], leading to the right from the East [Bent Street] entrance. If the Governor be there, the company is generally of a very brilliant description. There is a tolerable sprinkling of wealthy people and tradesmen, and also a liberal supply of fruiterers and lemon syrup sellers, and many others of the 'tag, rag and bobtail' species which are always attracted by a crowd.[104]

In 1902 Maiden stated there were no shelter pavilions in the Domain and that visitors were forced to retreat to the Art Gallery when it rained.[105] He was considering building a pavilion on Mrs Macquaries Point, using some of the old doric columns from the Darlinghurst Courthouse. The columns are probably the ones still used to support the statues in Centennial Park (see Chapter 8).

Swimming Pools and Bathing Houses

The Aborigines had fished and swum in the waters around Farm Cove for thousands of years. Last century bathing houses were established along the eastern side of Yurong (Ansons/Mrs Macquaries) Point — Robinsons and Fig-Tree being the best known.

By 1833 bathing had been made illegal in Sydney Cove.[106] Men had originally bathed nude but wore trunks as swimming gained in popularity. The Domain was a popular bathing spot because of the fine white sand (between rocks) which gradually eased into deep water, which was then as 'clear as crystal'. In 1837 it was reported that swimming had become so popular 'it was impossible to walk out at any time of day . . . without being annoyed by bathers in all directions . . . and a government order was issued that no person should be allowed to bathe openly . . . in the Domain . . . *after* six o'clock in the morning, or *before* six in the evening'.[107]

In 1850 there was 'a bathing establishment, in a retired part of the Promontory . . . from the heights [of which] the fair Australians may often be seen, in becoming costumes, stretching across the waters of the bay with all the agility of mermaids'.[108] Within five years several bathing houses had been built, with shark-proof netting or wooden grating. They charged for a bath and towel.

Much more popular was the free bathing place for men and boys called Fig-Tree, 'on account of a fine native tree of that description which overhangs the rock where the bathers undress themselves.[109]

There was also a threepenny segregated bath for women with a galvanised roof, on the Woolloomooloo side, being formed out of the hull of an old steamer with horizontal engines known as the *Ben Bolt*.[110] Its decks

View of Garden Island showing Bathing Houses on eastern side of Mrs Macquaries Point, 1865, *RBG Library*.

were divided into cubicles and the deckhouse fitted out to serve coffee, biscuits, brandy and cakes.

In 1879, at the time of the International Exhibition, four bathing houses can be identified — the Corporation, Robinsons (for gentlemen) and two additional bathing houses for ladies only. Robinsons bath houses had been established in 1837. The gentlemen's bath was from 3 to 30 feet deep, with 130 dressing rooms, while the ladies' bath, about a quarter of a mile distant, was 3 to 10 feet deep and had 40 retiring rooms. The Corporation Baths were more recent. They comprised two divisions, one of them free, while the other, established in 1861, was about the same depth as the gentlemen's bath at Robinsons, with 50 dressing rooms. The ladies' bath, established in the following year, was from 3 to 10 feet deep, and had 20 dressing rooms.[111]

In 1898 a flight of zigzag steps were built north of the Corporation Baths to link the high level walk with the 'beach-walk'. Prior to that the eastern side of the point had been much more inaccessible and 'mostly given up to loafers [and] now it would be an excellent rendezvous for respectable citizens'.[112]

Sometime in the early 1880s 'Professor' Frederick Cavill opened his elegant 'natatorium' in Farm Cove, near the Gardens. The natatorium was an open-slatted 'bath' supported by iron buoyancy tanks with dressing cubicles equipped with mirrors, brushes and combs, and it became one of the 'sights of the town'. The good professor advertised his facility with his own halting verse:

> Come everyone, each mother's son and every bonnie daughter,
> And learn to swim with sturdy limb and sport amid the water,
> For should you wish to swim like fish you have not far to travel,
> To Farm Cove go and soon you'll know the famous teacher Cavill.
> His bath secure, the water pure, no fear of monsters finny,
> For Cavill's there — his charge is fair
> He'll teach you for a guinea.

Frederick Cavill had six sons and they all swam. The fourth son, Arthur Roland 'Tums' Cavill is supposed to have invented the 'Australian crawl', but there was nothing new about the stroke. It had been used by indigenous American tribes, including the Aztecs, as well as in western Africa and on the islands of the South Pacific.[113] Indeed some have claimed that Aleck Wickham, a Solomon Islander living in Sydney, had given the Cavills the idea when they saw him swim at Bronte baths.[114] However, even if the Cavills didn't invent the stroke, they certainly promoted it and were responsible for its worldwide adoption.[115]

In 1900 the floating baths were moved from Farm Cove to the eastern side of Mrs Macquaries Point, opposite Garden Island.[116] And then, in 1909, the natatorium 'was swept away in a gale and deposited on a beach across the harbour, some considerable distance away'.[117]

In 1901 Parliament voted funds for the construction of baths in Woolloomooloo Bay, fronting onto the Domain.[118] But by that time the harbour was quite polluted: the city sewage had been emptied into it for years. The 'Bennelong Drain', a large brick-lined drain, still runs under the Tarpeian Precinct of the Domain to Bennelong Point. The drain has been altered several times, when the Opera House was built, and when the Opera House car park and Harbour Tunnel were constructed. It was part of an extensive drainage system constructed in the 1800s. One branch runs beneath Macquarie Street, another runs under the Intercontinental Hotel to the Bondi Outfall.[119]

From 1901 the waters of the harbour were becoming 'purer and purer' — the problem having been deferred by shifting the sewage to ocean outfalls!

J. H. Maiden was not happy with the bathing arrangements in Woolloomooloo Bay and Farm Cove in 1903. He reported that visitors to the Domain could see nude bathers from the road, and that the lessee of one of the baths had removed the 'scandal' by 'the erection of a hideous fence some 18 feet high'.[120]

Construction of the Government Bath commenced in 1903, at a cost of over £3,000, on the site of the old George Farmers Baths and Robinsons Hot and Cold Sea Water Baths, and was handed over to the Sydney Municipal Council on 30 September 1907; the Council paying the Government £1,850 for the site.

In November 1907 the City Council commenced the construction of new municipal baths on the site of the old ones. They boasted 250 dressing boxes, galleries with seats for up to 1,600 people, a diving tower 54 feet high. The pool measured 110 × 35⅓ yards and was 25 feet deep; there was a children's pool, 70 × 30 feet, on the site of the old Fig-Tree Free Baths. Earth-closets were also constructed, to be used until such time as they could be connected to the sewerage system.[121]

Governor's Bathing House, *RBG Library*.

The less than attractive Municipal Baths were opened to the public on 17 October 1908, at a cost of £11,149, and were lit inside and outside by electricity for the first time for evening carnivals (at a time when the rest of the Domain was still lit by gas lamps).

The old Corporation Baths, occupied and conducted for many years by Mr Richard Hellings (and known as Hellings Baths) were also demolished to make room for the new ones, while the Government Swimming Bath for men (a 60 × 40 foot pool controlled by the Gardens, built in 1903 at a cost of about £3,000) was also transferred to the Sydney Municipal Council in 1908 and became a Ladies Swimming Bath.[122]

A new boathouse was also built by the Fire Brigade Rowing Club beside the Ladies Baths, adjacent to the boat shed of a Mr Press.

In 1924 Andrew 'Boy' Charlton set three world records in the 200, 400 and 800 metres, swimming in the Domain Baths against the then world champion, Arne Borg of Sweden. The name of current baths still commemorates the name of 'Boy' Charlton.

A century earlier, in about 1826, Governor Darling had constructed his own private 'Gothick' Bathing House, with castellated walls and octagonal towers of stone found on the spot down from Government House on the western side of Farm Cove. The cost for materials was reported as being £22 0*s*. 5*d*.; the total expense, including 2,500 days of convict labour at 1*s*. 3*d*. per day came to £704 0*s*. 3*d*. It was demolished in June 1924, the stone being used for the construction of a new fernery, opened in December of that year.[123]

Children's Gymnasium

In 1904 a children's gymnasium was established in the Phillip Precinct of the Domain where the chess table stood until c. 1988. It was very popular with the children and crowded in school holidays on fine afternoons and as 'lively as a beehive'; and 'during the 12 months only 3 accidents (had) been recorded. These were at once taken to the hospital and attended to, and in a short time the patients were back at the swings again.'

In 1905 'a pleasing addition' — a substitute 'merry-go-round' in the form of a 'wheel-push-round' was made:

> An old cartwheel was set horizontally on its own axle and fitted perpendicularly into a solid block of stone sunk into the ground. The spokes and rim were then boarded over to a width of 8' 6'' diameter, forming a flat revolving platform without seats. On this disc the youngsters sit, or lie or roll and there are always some volunteers to do the pushing business. I have frequently seen 20 to 25 children, babies and all, on this simple piece of apparatus and as it is raised only 15'' from the ground and there is plenty of clean sand to fall upon, there have been no accidents during the six months of its existence.[124]

The old gymnasium no longer exists but was replaced by the one on the opposite side of Art Gallery Road.

The Aviary, Pergola, Insectarium and Zoo

The first pergola in the Botanic Gardens was constructed at the eastern end of the Middle Garden (area of the present Succulent Garden) before 1832, as indicated on the March 1832 plan.

On 22 April 1848, Charles Moore reported 'Having had the whole of the Arbour which was in this place, and in a most dilapidated state, removed'.[125] In 1849 Charles Moore again reported that 'The arbour was broken down from natural decay of the wood' and that it was removed, and 'the framework of a new one erected'.[126]

It is not well known that Australia's first Zoological Society housed its animals in the Gardens in this area.

Governor Macquarie had kept a few emus and kangaroos in the 'Domaine', while in 1833 a Bengal tiger was the star attraction in a cage at the corner of Hunter and Castlereagh Streets, and a limited number of birds and other animals were later displayed in Hyde Park. In 1851, the Yorkshireman Captain Charlesworth, brought his circus — consisting of an elephant, two bears, some red deer, a black sheep and a tiger — to Hyde Park.[127]

On 28 June 1852 an advertisement in *The Empire* called a public meeting, 'To take into consideration the desirableness of establishing a Zoological institution in or near the city of Sydney, for the encouragement of Science, and the recreation of the public'.[128] Following this meeting a committee was set up to found a Zoological Society of New South Wales.

In 1854 some of the animals belonging to Captain Charlesworth were purchased and taken to the Sir Joseph Banks Hotel at Botany to attract customers. These animals were later moved to the Marine Hotel of Hank Billings at Watsons Bay, where thousands flocked to see the trainer put his head in the lion's mouth. When Billings died the State rejected his widow's plan that the State take over the animals. It is believed they were poisoned.[129]

An early reference to an aviary in the Botanic Gardens comes from a lithograph of S. T. Gill, published by Allan and Wigley in 1856.[130] A substantial aviary was constructed adjacent to the pergola in 1860 and on 22 July the *Sydney Morning Herald* enthused:

> The Aviary in the Botanic Gardens — most of the visitors to the Sydney Botanic Gardens expressed their admiration of the beauty of the trellised enclosure along the eastern boundary of the Upper Garden walk, being for about a length of one hundred yards encircled by roses, woodbine, passion flowers and other popular varieties of climbing plants, which are trained round the trellis work and clustered in graceful profusion. The attractiveness will now be further increased by the construction of an aviary in the centre of the walk.[131]

The aviary was apparently built to house the donation of a collection of birds by Alfred Denison, brother of the governor. Additions were made to this collection from time to time and more cages had to be constructed.

> The building erected by the Government in the Botanic Gardens to provide for the collection of birds presented by Mr Alfred Denison to the Colony, is now open for public inspection. The structure is of wood upon stone foundations, and is sixty-four feet long and divided into eight compartments, the sides and tops of which are formed of with wire work. The top of the aviary rises towards the back, thus affording a succession of ridges, the elaborate decorations of which, together with the chaste alternations of green and white colours produce an extremely cheerful and pleasing effect. The crockets on the upper ridges are fleurs de lis, and the cornice is ornamented with a larger but equally tasteful design. At each of these is a neat finial surmounting ornamental projecting gables with pendants.

The Aviary, Botanic Gardens, 1862, showing symmetrical fountains and Small Lion in foreground, *RBG Library.*

> The cages are liberally furnished with perches and roots and the ground is sown with grass . . . Cast iron railings are being fixed in front of the aviary, at each end of which are neat stone fountains, that will constantly play, and will, from their overflow, supply the smaller fonts in the cages.[132]

The first non-avians to be added to the collection were an anteater and a Chinese deer in 1862.

The Acclimatisation Society of New South Wales had been set up in 1861. The President of the Society was his Excellency Sir John Young, Bart., KCB; Vice-President, Sir William McArthur; Council, Dr G. Bennett, Mr J. B. Bossley, Hon. J. Byrnes, MLA, Mr Charles Campbell, Dr Cox, Mr R. Fitzgerald, Mr W. Hemming, Mr J. F. Josephson, MLA, Dr Moffitt, Hon. J. B. Wilson, MLA, Mr Windeyer, MLA, Rev. R. W. Young; Honorary Secretary, Dr G. Bennett; Secretary and Treasurer, Mr William Thomas.[133]

In 1864 the Acclimatisation Society erected cages in the Gardens to receive the birds, they were kept 'in a very neat aviary, under the management of the bailiff'. The most noteworthy were the stately Secretary birds, male and female, of Southern Africa, and two species of Curassow from Brazil.

> The breeding stock of Pheasants in the aviary at present consists of eight distinct and valuable species, viz., the Silver, the Golden, the Ring-necked, the Sommerring or Japanese, the Roleege, the Matchurim, and the Common Pheasants. Endeavours have been made to procure several other rare kinds not as yet in the Society's collection.[134]

Other birds bred included English blackbirds and thrushes, which were later released in the Gardens to breed. It is interesting to note in this context that the English blackbird still breeds in these Gardens, while being less common in other parts of Sydney.

In 1867 'A few strange animals [were] found boxed off in corners, contented and seemingly happy as if they were not thousands of miles away from their native haunts'.[135] Eleven years later another trellis was erected on both sides of the walk in front of the aviary.[136] In 1915 this was replaced with

> A solid hardwood and galvanized-iron pipe pergola, in the Italian style, has been constructed to replace the old wooden framework, which it was necessary to demolish owing to decay and danger to the public. The old creepers were carefully preserved, and soon clothed the new structure.[137]

In 1880 the sum of £100 was voted for a monkey

The Bird Walk, 1890, *RBG Library*.

house. One of the baboons escaped one day and peered into the window of the cottage of the head gardener, terrifying his wife, who had the presence of mind to pull down the blind. The keepers later caught the baboon and returned it to its cage.[138]

The monkeys proved to be a popular attraction, but not for long, for in 1883 most of the animals, including some of the birds, were handed over to the recently formed Zoological Gardens at Moore Park.[139] Some of the birds were then returned to the Botanic Gardens because of public demand. The old stone sheds of the Aviary were dismantled in 1902 and replaced with new structures which were not finally removed until about 1940.

A former path in the east of the Succulent Garden was long known as the Bird Walk, and the foundations of some of the aviaries were uncovered when the Succulent Garden was being constructed.

In 1910, an Insectarium was 'erected in one of the enclosures adjoining the aviaries in which . . . the life

Insectarium at end of Aviary Pergola, 1915, *RBG Library*.

Crowds attending International Exhibition at Garden Palace, *RBG Library*.

histories of plant pests is being studied and demonstrated under the direction of the Government Entomologist'.[140] It was on the left hand side at the northern end of the pathway through the old zoo, and was probably demolished at the same time as the aviaries. The trellis had been pulled down around 1936.

There were at least three tortoises in the Gardens, 'kept near the offices by the flowerbeds' between the Anderson and Cunningham Buildings.[141] One of them may have come to the Gardens via the Solomon Islands.[142] The last living representative of the zoo was a large female tortoise. Although she wandered freely she kept to the dense shrubberies and, despite her size, was rarely seen. She died about 1967. Stuffed, she is now on display in the entrance foyer of the Brown Building.[143]

The Garden Palace

The Garden Palace, officially opened in Sydney on 17 October 1879, by the Governor, Lord Loftus, was the wonder and experience of a whole generation. The magnificent structure, designed by James Barnet, Colonial Architect, was built to house the Great International Exhibition of Sydney — a Trade Fair or Expo of its time — designed to promote Australia and Australian products and workmanship. The concept was derived from the Great Exhibition of Works of Industry, held in London in 1851 in the Crystal Palace — a new development in architecture, using metal and glass, designed by Joseph Paxton. Paxton is reputed to have obtained his inspiration for the load-bearing metal structures for his palace from the veins of the giant water lily, *Victoria amazonica*, which first flowered in cultivation at Chatsworth House in 1849.

Such exhibitions were seen to bring both profit and prestige, as well as upholding the sureties of Victorian England and the Brave New Industrial Age. Work commenced in 1878 on the building which was housed in the Inner Domain, adjacent to the old stables that were to become the Conservatorium of Music. It was erected in a remarkably short time, using electric lights for the first time to allow the construction to be carried on at night.

The completed building was 244 metres long and

Interior Dome of Garden Palace, *Government Printing Office.*

stretched from the site of the Conservatorium to the region of the State Library of New South Wales, with a central dome 30 metres in diameter and 64 metres high, and cost of £191,800. Old photographs indicate that there was also an obelisk constructed between the Palace and the road.

The exhibition was made up of primary products (including food, wine and farm animals), forest resources (timber, gums, resins and dyes), mining displays, manufactured goods (sugar, wool) and Australian-designed agricultural machinery, and continued until 20 April 1880, when the awards were made. A total of 1,117,536 patrons visited the exhibition, as counted on the newly patented turnstiles. After the closure of the exhibition the Macquarie Street entrance to the Garden Palace was besieged by a crowd of labourers seeking employment in the packing and removal of goods.[144]

The Palace, built of timber and corrugated iron cladding on brick foundations, was initially intended as a temporary structure and would have been difficult to maintain. After the exhibition the great halls were used for public meetings, concerts and balls. The galleries were converted into displays of the Technological, Industrial and Sanitary Museum and the Department of Mines installed permanent geological exhibitions. The basement was used to store the government archives, including the returns of the 1881 census.[145]

On 22 September 1882, the building was totally destroyed by fire. Arson was suspected and there were two unsubstantiated rumours at the time. One was that residents of Macquarie Street resented the massive intrusion on their view; the other was that details of the 1881 census (destroyed in the fire) may have been used to reveal the convict taint in some families aspiring for social respectability.[146] Also lost in the fire were all land-occupancy records, archival records, the colonial collection of statuary, the paintings of the Art Society of New South Wales, and the library of the Linnean Society.

However, the loss of the Palace was a gain for the Gardens, as the grounds of the building were gradually incorporated into the Botanic Gardens and became known as the Palace Garden.

In 1889 the main entrance gates opposite the Public Library (State Library of New South Wales) were completed, the central metal section of the gate representing the dome of the Palace Garden. This gate was later moved to Macquarie Street when the Cahill Expressway was being constructed.

During the sesquicentenary of European settlement in Australia in 1938, the Sunken Garden was built in memory of the pioneer men and women at the spot where the central dome of the Palace had been. A plaque was later laid at this site to commemorate the Garden Palace.

When the International Exhibition closed, Joseph Henry Maiden was appointed curator of the Technology Museum (Technological, Industrial and Sanitary Museum) — later to become the Museum of Applied Arts and Sciences. After the fire in 1882 the Museum was located in a large galvanised-iron building in the Phillip Precinct of the Domain, and moved to Harris Street, Ultimo, in 1893. The galvanised-iron building was then used to house the Mining and Geological Museum.[147] This shed, better known as the Mining Museum, and the Government Architect's workshops near St Mary's Church, were not removed until 1914.

The shed had been used as a temporary horticultural annex to the International Exhibition of 1879–80, and afterwards used a temporary shelter for the unfortunate Italians of the disastrous expedition of the Marquis de

Rays to New Ireland.[148] The Italian colonists, wishing to settle in New Ireland, had sailed from Barcelona in July 1880. Eleven died on the voyage and a further 22 died on arrival at New Ireland. In December 1880, they set out for Australia. After troubles in Noumea, the New South Wales government came to their assistance.[149] After their sojourn in the Domain a settlement called 'New Italy' was established near Woodburn on the Richmond River.

The extraordinary demand for space in the main building of the Garden Palace had necessitated the erection of a plethora of other exhibition stalls in the Domain. One of these was the Fine Arts Gallery (1879–85) to the west of the main entrance to the Botanic Gardens, which was the precursor of the Art Gallery of New South Wales. As well as this there were two matching annexes (Machinery Halls) in the Domain, each 300 feet × 160 feet, constructed by Mr John Young for a cost of £16,000. They were rather functional in design but their surroundings were pleasant, consisting of lawns and avenues of well-established ornamental trees.[150]

An inspection of a map of the time gives a more detailed list of the additional buildings and special areas associated with the International Exhibition:

1 Police Office
2 Photographic Gallery
3 Three ticket offices (one of these wooden structures was still standing 'near the native border' in 1903)
4 American Novelty Company
5 Popcorn stand
6 Wallerawang Coal and Iron Company
7 British and Foreign Bible Society
8 Japanese Kiosk
9 Turkish Bazaar
10 (Howell) Confectionery stand
11 (Cripps) Restaurant (in the north-east corner of Lawn 8 below the Arc and the Pyramid)
12 Sherwood Tile Company (Schweppes Mineral Water)
13 Australian Dairy
14 German Pavilion
15 (Emerson) Oyster Saloon
16 American Coltage Company (Refreshments)
17 Kingdom and Kelman (Refreshments)
18 Monumental and Statuary Exhibit (Hanson)
19 French Commission
20 (Drysdale and Robert) Machinery Hall
21 Austro-Hungarian Wine and Beer Tasting Hall
22 Fiji House
23 Concrete Cottage
24 Restaurant (Cotton)

Palace Garden after fire with Huntsman and Dogs Statue in foreground, *Government Printing Office.*

Summer House in Palace Garden, 1881, *RBG Library*.

25 Judges Office
26 (Compagnone) Restaurant
27 (Compagnone) Restaurant
28 Camera Obscura
29 (Caster) Confectionery
30 Maori House (in Gardens at site of Pyramid Glasshouse)
31 Horse and Cattle Parade Ground
33 Austrian Refreshment Pavilion
34 French Cafe

Other buildings associated with the Exhibition included the Garden Palace (Palace Garden), two Machinery Halls (with tramway; Phillip Precinct Domain), the Agriculture Hall (Phillip Precinct Domain), and the Fine Arts Gallery (Upper Garden). There was also a block of toilets (latrines) in the region of the Depot.

The map also shows:

Guard House and Gates on Bridge Street
Lodge and Gate on west of Farm Cove
Fence around Government House Grounds
Summer Houses in west and east of Botanic Gardens
Four bathing houses — Corporation, Robinsons (gentlemen), and two ladies baths
St Marys Lodge and Gate
Part of Macquarie Wall
Palmer Street Gate
Line of Roadway in Domain
Completed Farm Cove Reclamation

Cricket Pitch and Bowling Green

The Phillip Precinct of the Outer Domain was used for many years by the New South Wales Cricket Association. The first Intercolonial Match (New South Wales *vs* Victoria) was held in Melbourne in 1856 (won by New South Wales by one run) and in December that year a cricket site was set apart in the Domain in lieu of the one that existed at Hyde Park.[152] The return match was held in the Sydney Domain in 1857 (and also won by New South Wales).[153]

The first representative match between England and New South Wales was played in the Domain in 1862.[154] The atmosphere of this match was later evoked in an article in the *Sun*:

> The official marquee was erected on the top of the western slope. Various side attractions in the way of booths were on the opposite side of the oval. There was no scoring board, but, instead a tent, containing a small printing plant . . . from which, for a small charge, an octavo card giving the names and scores of the players [was] issued.[155]

In 1907 the little humpy at the edge of the Cricket Ground was removed, and a small neat pavilion for cricketers and footballers was erected near the bridge (over Botanic Gardens Creek) with toilets and showers, and with storage room for lawn mowers, horse-feed and small tools.

At the same time a brick toilet for Domain workmen was built in the corner of the work yard for £10.[156]

In comparatively recent times there was a Bowling Green and clubhouse on the western side of the Phillip Precinct of the Domain, with a low fence around it, called the Parliament House bowling green in 1914.[157] A plaque, dated 1961, and kept in the Gardens Library, is inscribed as follows:

> Ray Carroll Green
> To the memory of the late W. R. (Ray) Carroll Founder and Patron of the
> Domain Veterans Bowling Club.

This Bowling Green was removed in the early 1970s, about the time the old brick change sheds (near the air vents to the eastern suburbs railway line) were demolished and replaced with new change sheds near the ladies toilets.[158]

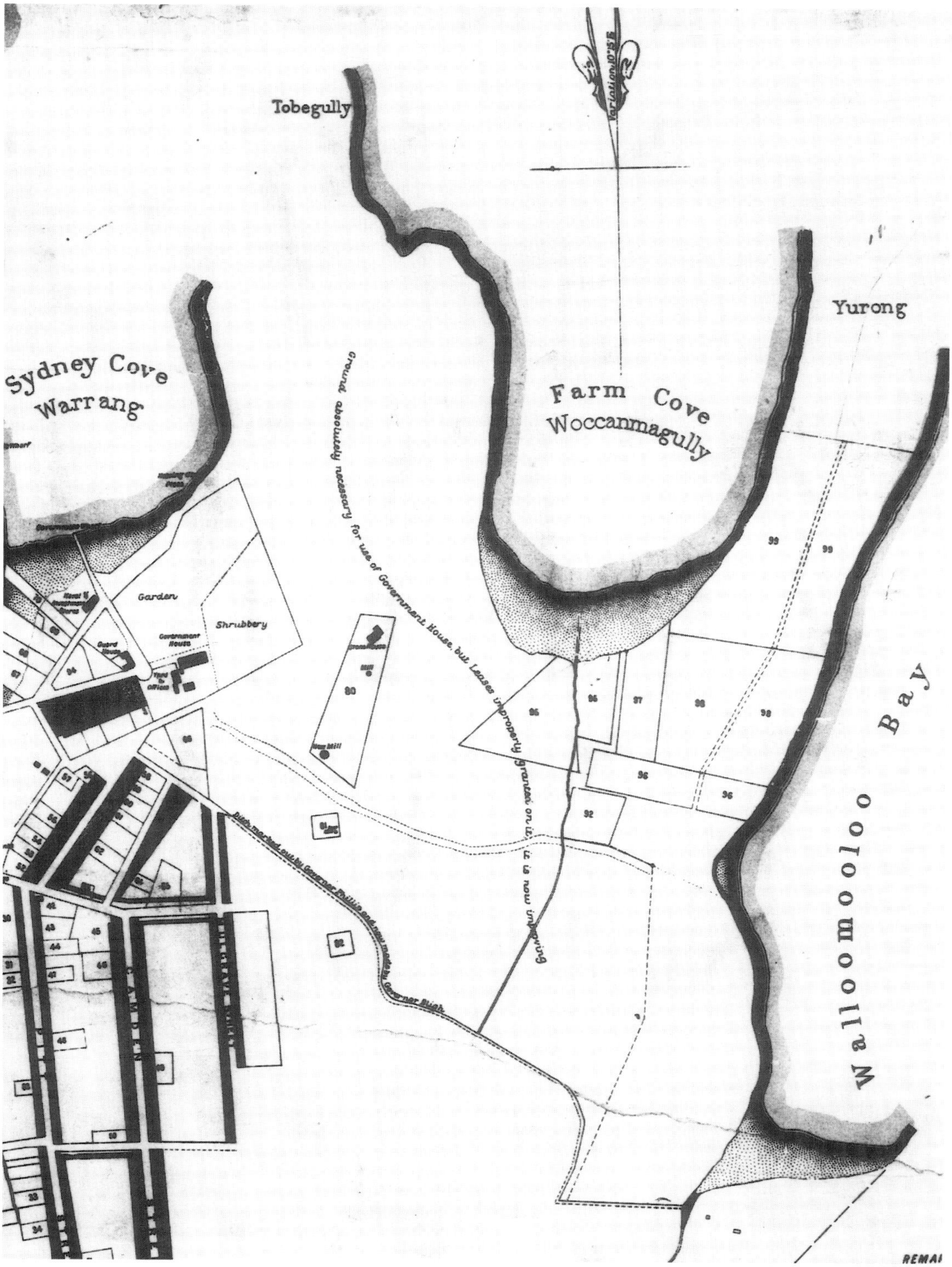

Map 1. Portion of a 'Plan of the Town of Sydney' by James Meehan showing the 'Governor's Domain' as it was in 1807, *Central Mapping Authority of NSW.*

2

ROADS AND GATES

Mrs Macquaries Road and Chair

From Surgeon Harris' letter of 25 October 1807 we learn that there were already 'Carriage roads all round Bennelong's Point, and down about "Farm Cove" ', and these were 'some time since constructed'. But when the energetic Governor Macquarie arrived in 1810, he set about developing an extensive network of roads, expanding on those laid out by Bligh and earlier governors.

While Fort Macquarie was being built (1817-19 — at Bennelong Point, where the Opera House now stands) much of the point was cut away to accommodate the road from Sydney Cove (Circular Quay) to Farm Cove. The road at the point went under the so-called Tarpeian Way — widened again in 1917 during roadworks in Macquarie Street[1] — the Tarpeian Rock being a cliff from which ancient Roman criminals were hurled.[2]

In 1813 Macquarie commenced a road through the Domain for his wife's recreation. Called 'Mrs Macquarie's Road', it was completed on 13 June 1816. A bridge would have had to be built over Botanic Gardens Creek at the time when the road was constructed. The brick culvert just south east of the Gardens Restaurant — dubbed the 'Macquarie Culvert' by Gardens' staff — may well date from this period. If so it would make it the oldest existing bridge in Australia — nine years older than Tasmania's Richmond bridge.

As Mrs Macquarie liked to sit at Ansons Point (later called Mrs Macquaries Point) and take the sun, the Governor had workmen carve a special seat for her (Mrs Macquaries Chair) out of a rock ledge, which remains as a noteable landmark. The date of completion of the road is recorded above the chair as follows:

> Be it thus Recorded that the road around the inside of the Government Domain, called Mrs Macquarie's Road, so named by the Governor on account of her having originally planned it, Measuring 3 miles and 377 yards, was finally completed on the 13th day of June 1816.

The traditional foundation date of the Botanic Gardens is taken as this date.

On this 'auspicious' day Governor Macquarie noted that at 1 p.m. 'Nicholas Delaney, the Overseer of the Working Gang employed ... in the Government Domain reported to me that Mrs Macquarie's New Road [was] completely finished' and Mr Delaney and his gang of 10 men were provided with five gallons of spirits with which to celebrate this achievement.[3]

Fifteen years later, on 20 August 1831, Charles Fraser informed the Colonial Secretary that 'the roads in the Government Domain are now made passable for two carriages'. On 9 September, he states that 'the Domain will be opened for carriages on Tuesday next, the 13th inst. [while] the road from the stairs near Fort Macquarie along Farm Cove to the gate which crosses the road at the extremity of the Botanic Garden is reserved for the exclusive accommodation of persons on foot'. He then went on to say that 'Carriages and horsemen may enter the Domain at the gate near the School of Industry, or at the Woolloomooloo Gate [Palmer Street Gate] at the southern boundary of the Domain'.[4]

This was essentially the birth of the Domain as we know it today.

In 1846 'great improvements [were] made [to] the approaches [to the Gardens], by means of new fencing, new gates, a widening of the road, etc., so as to admit of the more easy approach of carriages — and others of the same kind'.[5]

Parts of Mrs Macquaries Road still follow the same pathway today (1988) as they did in Cartwright's map of 1816. For example, the pathway north of the Macquarie Wall in the Gardens (where the Swamp

Mrs Macquaries Chair, 1989, *Jaime Plaza.*

Mahoganys grow and the road at the back of Sydney Hospital (where oak trees had been planted in Macquarie's time[6]) follow the route of the old carriage road.

In 1902 J. H. Maiden reported that 'A sign of the times is the large number of these horseless vehicles now to be seen in the Domain. They do not, however, appear to have any effect on the horse traffic.'[7]

Restricted Access (1816)

Initially there was restricted access to the Gardens, which were not fully opened to the general public until 1831 when both the Gardens and Domain were closed and locked at night.[8] Louisa Meredith, writing of Sydney in 1839, said 'It was unfashionable either to walk or drive in the Domain [as it] was a notorious fact that maid servants and their sweethearts resorted thither on Sundays and, of course, that shocking circumstance ruined its character as a place for their mistresses'.[9]

There may well have been another reason why the Domain was not popular with ladies. No ladies toilets were even considered for the Domain before 1905![10]

Old Gateway to School of Industry in the Macquarie Wall in the Domain, 1910, *RBG Library.*

'The Park where the Gates Never Close' (1860)

The Domain (St Marys) Lodge was built in 1835 at the gates at the southern end of what is now Art Gallery Road.[11] An 1845 drawing of the gates by Conrad Martens shows four piers (with lights on top of the inner pair) and wrought-iron gates. A new gate was constructed in 1858 by A. Dawson.

In 1867 a pair of handsome gates had been set up at the Palmer-street entrance to the Domain, and a pair of equally handsome gates had been put up at St Marys Lodge.[12] The existing sandstone pillars at Domain Lodge are dated 1876.

Meanwhile, from 1860, under the administration of Lands Minister Robertson, the Domain remained open at night and was to become 'The Park where the gates

Bent Street Entrance to the Domain, 1892. Note the Palace Garden Gate in its original position, before it was moved to Macquarie Street. The Bourke Statue, flanked by the two cannon (now in Centennial Park) occupied the current site of the Shakespeare Memorial, *RBG Library.*

Main Entrance to Botanic Gardens, 1856, *RBG Library*.

never close', although carriage traffic remained restricted after dusk for many years.[13]

The gate and stile at the east end of the Bent Street entrance offered free admission from 1816. The earliest photograph of this site, showing ornamental gates, has been dated at 1842. It is not known when these gates were constructed. They were moved in 1912, during the widening of Macquarie Street, to the top of Fig Tree Road.[14]

The 'Principal' Entrance (1873)

The main entrance, to the Botanic Gardens — Woolloomooloo Gate — was opposite the stone drinking fountain in the Outer Domain. The position of this gate may be established by standing in Central Avenue in the Phillip Precinct of the Domain, looking at the statue of Sir John Robertson and imagining the continuity of the avenue in a straight line.

In the early days, a road led through the Upper Garden (contained on either side by a paling fence shaded by trees).[15] This pathway, running essentially north-south, near the Levy Fountain, is not to be confused with 'Fig Tree Avenue' (planted 1847) which ran essentially east-west from the site of what is now the Shakespeare Statue Group to where the stone drinking fountain used to be near the gate.

The public entered the Middle Garden by walking down the pathway (contained by the paling fence and an arcade of acacias) through an old wooden gate, painted to represent granite,[16] to the two Cabbage Palms, *Livistona chinensis* (planted 1840) which marked the start of the Gardens:

> At the entrance to this avenue, on the right hand side, there was later a notice board; on this were painted the Regulations of 1848. The regulations, referring to the opening of this gate on Sunday until one o'clock, were observed in the following manner: His Excellency the Governor and suite attended service at St James' Church. Usually they walked back across the Domain. At the locked-

entrance gate stood a regular attendant and a group of waiting citizens. The official, having removed his hat, unlocked the gate. His Excellency and party entered; a few minutes later the citizens followed. The Governor walked down the road, then turned to the left, and entered Government House grounds. The citizens would go straight on to the Middle Garden.

Those who did not go to church had to content themselves in the Outer Domain, swimming the dog, having a bath at the Fig Tree or some other part of Woolloomooloo Bay, or by wandering about the Domain, then covered with a dense scrub consisting of 'tickbush' (*Kunzea corifolia*).[17]

The original timber gate was replaced by an ornamental sandstone and wrought-iron gate. Designed by James Barnet, Colonial Architect, it was erected at the eastern end of Fig Tree Avenue near the stone fountain in the Domain in 1872–73 (1873 recorded on the stone) at a cost of £408. The wrought iron gates came from Robert Dunlop's workshop in Forbes Street, Woolloomooloo, (costing £145) and the carved sandstone piers came from Messrs Hanson and Sharp.[18] It is an elegant composition of four sandstone piers, with flanking walls and graceful wrought-iron gates. The piers are finely carved with floral motifs and are surmounted by sandstone urns.

This gate, now called the Woolloomooloo Gate, was moved to its current site in 1962, during the construction of the Cahill Expressway. The ornamental pools were then built in 1966 for £325 ($650).[19]

The Yurong Gate (1837, 1901)

In 1833, when Richard Cunningham, the then Superintendent (Director) of the Gardens, was laying out the Lower Garden (garden north of Macquarie Wall), he wrote a letter requesting that the road from Sydney Cove to Yurong Point currently dividing the Government Domain (Mrs Macquaries Road) proceed along the beach of Farm Cove.[20] The Yurong Gate was established in 1837. 'Yurong' was the Aboriginal name for the point — also known as Ansons Point, after John Anson, who had a farm there, and now called Mrs Macquaries Point. The current gate was constructed in 1901.

Main Entrance to Botanic Gardens from inside Gardens after the completion of the new gates (Woolloomooloo Gate) in 1873. These gates were removed to their current site during the construction of the Cahill Expressway, *RBG Library*.

Government House Gate (1869)

The site for New Government House was chosen in 1827. Work did not begin until 1837. The Lodge was constructed in the early 1840s, but Government House was not occupied until 1845. The sandstone and wrought-iron gates and associated fencing designed by James Barnet, Colonial Architect, were constructed in 1869. Government House Delivery and Inner Domain Depot Gates were probably constructed at the same time.

The lodge and gates were moved to the east in 1916 because of the widening of Macquarie Street. The gates were moved again in 1956 to the eastern side of the lodge, as part of the work associated with the construction of the Cahill Expressway.

The well-proportioned entrance consists of five sandstone pillars and attractive wrought iron gates which show the emblems of the shamrock, thistle, rose and crown, and provides for vehicle access to Government House.

Victoria Lodge Gate (1864)

A pair of dated stone piers was erected at the Victoria Lodge Gate site in 1864. These were replaced in 1901 and the piers were moved to their current site in the Macquarie Wall, near Lion Gate Lodge. The original metal gates from Victoria Lodge were renovated and installed at Lion Gate Lodge in 1989.

Palace Garden Gate (1889)

The original gates to the Garden Palace were constructed in 1879 in association with the Great International Exhibition. After the destruction of the Garden Palace by fire in 1882, its grounds were gradually incorporated into the Gardens. The Palace Garden Gate was erected at the Bent Street corner, where the Morshead Gate now stands, opposite the Mitchell Library in 1889 (a year later than the date on its base), at a cost of £1,033 7*s*. 6*d*.

It consists of four sandstone piers, topped with wrought-iron finials and flanked by sandstone walls. The gates are of very elaborate wrought iron, and are surmounted with a representation of the dome of the Garden Palace.[21]

The gate was moved some 60-70 metres to the north when Macquarie Street was widened in 1912 and then relocated to the east in 1924/26 as part of the formation of Shakespeare Place.[22] It was again moved in 1959 when the Cahill Expressway was being constructed and erected in its present position (Macquarie Street) in 1962. J. H. Maiden noted that:

Place Garden Gate, 1989, *Jaime Plaza*.

> The gates are a fine example of artistic work in iron and stone. The structure is composed of three gateways, viz. a centre with double gates, and a smaller gate on either side. Above the horizontal bar at the top of the central gates there is a representation of the dome of the Garden Palace.
>
> The pillars are of Waverley sandstone, and give one the impression of massiveness and strength. The carving is rich, interspersed with the ordinary floral subject, representing some of our native plants: the 'Flannel Flower', 'Waratah', 'Flame Tree', 'Firewheel Tree' and Banksia. On the main pillars are two very fine medallions in bronze.[23]

Morshead (Shakespeare) Gate (1963)

After the relocation of the Palace Garden Gate, a new gate was constructed opposite the Mitchell Library in 1963 and called the Morshead Gate. The entrance name was changed to the Shakespeare Gate in 1984 as the area had earlier been called Shakespeare Place — relating to the monument to William Shakespeare — and renamed the Morshead Gate in 1990.

The adjacent fountain, unveiled by Sir Roden Cutler on 27 January 1966, also honours the memory of Lieutenant General Sir Leslie Morshead, and all who served with him in two world wars.[24]

Queen Elizabeth II (Man-o'-War) Gate (*c*. 1870, 1954)

This Gate, originally built *c*. 1870, was given its regal name after the Royal visit of 1954 — the first time a reigning monarch had visited Australia. In 1959 the Gardens acquired the 'Royal' prefix.

It was previously called Man-o'-War Entrance because sailing ships of the British and foreign navies used to moor in Farm Cove (sometimes called Man-o'-War Bay), where they were protected 'from the prevailing westerly gale in winter, or the occasional southerly squall in summertime.[25]

Following the Queen's visit during her Silver Jubilee in 1977, a bronze plaque was unveiled at this entrance, commemorating what was to be called the 'Jubilee Walk' around Farm Cove.

Henry Lawson Gate (1968)

This gate was named after the statue of the writer Henry Lawson (1867–1922) which was unveiled in the adjacent precinct of the Domain in 1931. In 1962 Miss Jill Hellyer, Honorary Secretary of the Fellowship of Australia Writers, wrote to the Minister for Agriculture, complaining that the 'gates opposite the Henry Lawson memorial in the Domain ... are continually closed [with] no path cleared to them'.[26]

Opera House (Alexandra) Gate (1973)

From old maps it can be established that prior to 1897 the Man-o'-War (now Queen Elizabeth II Gate) was the only entrance in the area of the old fort where the Opera House now stands. In 1897 about six acres from the Inner Domain was added to the Gardens (between Government House and the pathway west of the Twin Ponds) including the area near the Opera House Gate.

This Gate, previously called the Alexandra or Governor's Private Gate[27] was renamed after the Opera House when it opened in 1973.

(New) Rose Garden and Herb Garden (Conservatorium) Gates (1988)

Two gates were established at the rear of the Conservatorium in 1919, but were closed during the construction of the underground railway. They were replaced by the Rose Garden Gate and the Conservatorium Gate in 1933/34.

The old Conservatorium Gate, was renamed the Rose Garden Gate in 1988 because of the opening of the (New) Rose Garden on Lawn 15. Consequently the Old Rose Garden Gate was called the Conservatorium Gate. In 1990 this was changed to Herb Garden Gate to commemorate the new garden to be developed there.

Chronological Sequence of Gates as Opened/Named/Built

1816 Macquarie Wall and gateway and style at eastern end of Bent Street — access limited. Bent Street Gate moved to top of Fig Tree Avenue in 1912.

1831 Main Entrance of enlarged and reorganised Gardens open to general public. Ornamental entrance erected 1873, moved to current site in 1962 during construction of Cahill Expressway — now called Woolloomooloo Gate.

1835 Domain Lodge — Gate Piers 1858.

1837 Yurong Gate requested by Richard Cunningham. The current gate was constructed in 1901.

1845 Domain (St Marys) Lodge Gate. New Gate constructed in 1858, and 1867. Existing sandstone pillars dated 1876.

1864 Victoria Lodge Gate — replaced in 1901. Dated stone piers 1864 moved to their current site in Macquarie Wall near Lion Gate Lodge in 1901. Actual gate renovated and replaced at Lion Gate in 1989.

c. 1867 Palmer Street Gate, replacing earlier gate.

1869 Government House Gate, moved to the east in 1916 because of widening of Macquarie Street, and again in 1956 with construction of the Cahill Expressway.

c. 1870 Man-o'-War Gate (Queen Elizabeth II Gate, 1954).

1873 Alexandra (Governor's Private) Gate (Opera House Gate, 1973).

1876 Old Domain Entrance Gates — date of existing sandstone piers at southern end of Art Gallery Road near Domain Lodge.

Gate in Macquarie Wall, *c.* 1880. Note the terracotta urn (currently on steps of Visitor Centre) and Garden Palace in background, *RBG Library*.

1889 Palace Garden Gate — moved to current site in Macquarie Street in 1959/62 during the construction of the Cahill Expressway. Now called Morshead (Shakespeare) Gate.

1901 Victoria Lodge Gate replaced earlier gate of 1864.

1901 Yurong Gate constructed (see 1837).

1919 Two gates established at rear of Conservatorium in 1919, closed for underground railway construction, replaced by Rose Garden and Conservatorium Gates in 1933/34.

1954 Queen Elizabeth II Gate named after the Royal Visit of 1954 — previously Man-o'-War Gate (see 1870).

1963 Morshead Fountain Gate (Shakespeare Gate, 1984–1990) constructed on original site of Palace Garden Gate.

1968 Henry Lawson Gate, named after statue unveiled in Domain in 1931.

c. 1973 Opera House Gate — named after the Sydney Opera House (opened 1973). Previously Alexandra Gate.

1988 Rose Garden Gate named after the New Rose Garden (previously called Conservatorium Gate). Consequently the Old Rose Garden Gate became Conservatorium Gate — changed to Herb Garden Gate in 1990.

3

THE WISHING TREE

The Original Vegetation[1]

It is understandable that the bush was alien to European eyes. Not everyone, however, reacted to it in quite this way. Captain Watkin Tench of the marines was lyrical in his descriptions of the local vegetation:

> The general face of the country is certainly pleasing ... covered for the most part with big spreading trees ... a variety of flowering shrubs abound, most of them entirely new to a European, and surpassing in beauty ... all I ever saw in an uncultivated state; among these, a tall shrub bearing an elegant white flower which smells like the English May.[2]

Tench goes on to say 'A few wild fruits are sometimes procured, among them is the small purple ''apple'' mentioned by Captain Cook'. He then perceptively notes his concern 'that [a] botanist was not sent out, for the purpose of collecting and describing the rare and beautiful plants with which the country abounds'.

Some idea of the natural vegetation of the Domain, as it would have appeared in the time of Governor Phillip, may be obtained by taking a ferry ride around Bradleys Head. To most Europeans it seemed a harsh dry environment with eucalypt woodland — 'brushwood' — and a variety of smaller plants occurring on exposed rocky outcrops and in the more sheltered fern gullies of the creek beds.

By 1816 the Domain had been largely cleared of its original forest growth, except for a narrow irregular strip of timber where Macquarie Street north now runs. There was also considerable regeneration. In Patrick White's novel, *Voss*, the explorer Voss (based on Ludwig Leichhardt) discusses his proposed expedition, *c.* 1847, with Frank Le Mesurier near the 'scrub and larger, ragged trees' above the water on the northern side of the Domain.[3]

In the Annual Report of the Botanic Gardens and Domain for the year 1902, the Director, J. H. Maiden, relates:

> The Outer Domain is of course the heart of the city of Sydney. People are alive who remember it to be a place of wild flowers, and they have spoken to people who remember it in much the same state as it was when the white men first set foot on its shores.

The wildflowers people saw would have included *Hibbertia scandens* with its yellow flowers; Acacias; the Woody Pear, *Xylomelum pyriforme*; Grevilleas and Banksias; the Blueberry Ash, *Elaeocarpus reticulatus*; Native Olive, *Notelaea ovata*; Christmas Bush, *Ceratopetalum gummiferum*; the Wonga Wonga Vine, *Pandorea pandorana*; and heath plants like the Native Fuchsias, *Epacris longiflora*. As well as these there would have been many pea flowered plants including bushpeas, *Pulenaea retusa* with yellow flowers, the red-flowered Running Postman, *Kennedya rubicunda*; and the purple *Hardenbergia violacea*. The Tongue Orchid, *Dendrobium linguiforme*, also grew on the rocks and produced delicate white flowers in spring.

Near the salt water, the vegetation included the Swamp Oak, *Casuarina glauca*; Port Jackson Fig, *Ficus rubiginosa*; and Cheese Tree, *Glochidion ferdinandi*. In the 1850s the undergrowth over a large area of the Domain consisted of various 'tea-trees' including species of *Leptospermum*, *Melaleuca* and *Kunzea* (tick bush) — probably regrowth.

The Cabbage Tree Palm, *Livistona australis*, grew in the valley of the Tank Stream and probably in the Domain, although they were not recorded by Maiden in his 1902 list. This palm was heavily exploited in the early days of the settlement for the construction of huts, for food, and for headwear.

The trees included the Smooth-barked Angophora;

Moreton Bay Fig trees on Mrs Macquaries Point, 1989, *Jaime Plaza*.

Angophora costata; the Swamp Mahogany, *Eucalyptus robusta*; the Bangalay, *Eucalyptus botryoides*; the Blackbutt, *Eucalyptus pilularis*; the Forest Red Gum, *Eucalyptus tereticornis*; the Red Mahogany, *Eucalyptus resinifera*; the Narrow-leaved Scribbly Gum, *Eucalyptus racemosa*; the Red Bloodwood, *Eucalyptus gummifera*; and the Turpentine, *Syncarpia glomulifera*. The first farm in Australia had been established on some alluvium to the east of the creek that still runs through the Botanic Gardens, but the sandy soil was too poor to support grain crops.

By 1870 the eucalyptus and banksias were fast disappearing from the Domain, due to natural decay and to white man's use of the land. However, some eucalypts on the eastern side of the Domain (along the present roadway to Mrs Macquaries Point) were still in good condition in 1879. Over the years the soils have been enriched to the extent that they are now too nutrient-rich to suit many of the original native plants.

In 1992 a few old warriors still remain in the Domain — notably gnarled old specimens of the Forest Red Gum. There are some younger Smooth-barked Angophoras and Red Bloodwoods which are presumably the product of natural regeneration, as well as the Bangalay and Blackbutt. Port Jackson Figs grow amongst the rocky outcrops east and west of the point, and there are still some Cheese Trees growing in similar areas, as well as a remnant *Clerodendron tomentosum* and *Rapanea variabilis* at Mrs Macquaries Point.

Maiden and Camfield's List of Native Plants in the Domain[4]

'List of plants Growing Without Cultivation in the Outer Domain 1902. In a few years the indigenous plants in the Domain will be far rarer than they are at present and for the information of the present generation and of posterity I have, with the very cordial assistance of Mr Julius H. Camfield, compiled a list of such species as are found growing in the Outer Domain in the year 1902.'

Name used	Currently accepted name	Name used	Currently accepted name
Dilleniaceae		Ficoideae	(Aizoaceae)
Hibbertia volubilis	*(Hibbertia scandens)*	*Tetragonia expansa*	*(Tetragonia tetragonioides)*
Pittosporeae	(Pittosporaceae)	Umbelliferae	(Apiaceae)
Pittosporum undulatum		*Hydrocotyle hirta*	*(Hydrocotyle peduncularis)*
Billardiera scandens		*Hydrocotyle asiatica*	*(Centella asiatica)*
Portulaceae	(Portulacaceae)	*Siebera billardieri*	*(Platysace lanceolata)*
Portulaca oleracea		*Apium leptophyllum*	
Hypericineae	(Hypericaceae)	Araliaceae	
Hypericum gramineum		*Panax sambucifolius*	*(Polyscias sambucifolia)*
Tiliaceae	(Elaeocarpaceae)	Loranthaceae	
Elaeocarpus cyaneus	*(Elaeocarpus reticulatus)*	*Loranthus celastroides*	*(Muellerina celastroides)*
Sapindaceae		Rubiaceae	
Dodonaea triquetra		*Opercularia aspera*	
Leguminosae	Fabaceae: Subfamily Faboideae)	*Opercularia diphylla*	
		Opercularia varia	
Viminaria denudata	(Viminaria juncea)	Compositae	(Asteraceae)
Daviesia acicularis		*Calotis cuneifolia*	
Pultenaea retusa		*Lagenophora solenogyne*	*(Lagenifera* sp. *A* var. *A)*
Glycine clandestina		*Lagenophora emphysopus*	*(Lagenifera* sp. *A* var. *B)*
Glycine tabacina		*Epaltes australis*	
Kennedya rubicunda		*Cotula australis*	
Hardenbergia monophylla	*(Hardenbergia violacea)*	*Cassinia longifolia*	
Leguminosae	(Fabaceae: Subfamily Mimosoideae)	*Cassinia quinquifaria* (sic.)	
		Gnaphalium luteo-album	*(Pseudognaphalium luteoalbum)*
Acacia juniperina	*(Acacia ulicifolia)*	*Gnaphalium japonicum*	Could be one of several species of *Gnaphalium* recognised today
Acacia suaveolens			
Acacia longifolia			
Acacia maidenii		*Senecio australis*	*(Senecio lautus)*
Acacia discolor	*(Acacia terminalis)*	Goodeniaceae	
Saxifrageae	(Cunoniaceae)	*Goodenia paniculata*	
Ceratopetalum gummiferum		Campanulaceae	(Lobeliaceae)
Crassulaceae		*Lobelia gracilis*	
Tillaea verticillaris	*(Crassula colorata)*	*Lobelia anceps*	*(Lobelia alata)*
Tillaea macrantha	*(Crassula macrantha)*	*Lobelia purpurascens*	*(Pratia purpurascens)*
Myrtaceae		*Wahlenbergia gracilis*	(Campanulaceae)
Leptospermum flavescens		Epacrideae	(Epacridaceae)
Kunzea corifolia	*(Kunzea ambigua)*	*Astroloma humifusum*	
Melaleuca ericifolia		*Monotoca elliptica*	
Angophora lanceolata	*(Angophora costata)*	Myrsineae	(Myrsinaceae)
Eucalyptus pilularis		*Myrsine variabilis*	*(Rapanea variabilis)*
Eucalyptus haemastoma	*(Eucalyptus racemosa)*	Jasmineae	(Oleaceae)
Eucalyptus robusta		*Notelaea ovata*	
Eucalyptus botryoides		Gentianeae	Gentianaceae
Eucalyptus tereticornis		*Centaurium spicatum*	
Eucalyptus resinifera		Convolvulaceae	
Eucalyptus corymbosa	*(Eucalyptus gummifera)*	*Polymeria calycina*	
Syncarpia laurifolia	*(Syncarpia glomulifera)*	*Dichondra repens*	
Lythrariea (sic.)	(Lythraceae)	Solaneae	(Solanaceae)
Lythrum hyssopifolia		*Solanum verbascifolium*	*(Solanum erianthum)*

Name used	Currently accepted name	Name used	Currently accepted name
Bignoniaceae		*(Juncus planifolius)*	(Juncaceae)
Tecoma australis	*(Pandorea pandorana)*	*Juncus caespiticius (sic)*	
Myoporineae	(Myoporaceae)	*Juncus bufonius*	
Myoporum acuminatum		*Juncus homalocaulis*	
Verbenaceae		*Juncus communis*	*(Juncus continuus)*
Verbena officinalis		*Juncus pauciflorus*	*(Juncus usitatus)*
Clerodendron tomentosum		*Juncus prismatocarpus*	
Labiatae	(Lamiaceae)	Cyperaceae	
Plectranthus parviflorus		*Cyperus polystachios*	*(Cyperus polystachyos)*
Chenopodiaceae		*Cyperus tenellus*	
Chenopodium carinatum		*Cyperus enervis*	
Amarantaceae	(Amaranthaceae)	*Fimbristylis diphylla*	*(Fimbristylis dichotoma)*
Amaranthus		*Scirpus riparius*	*(Isolepis cernua)*
Alternanthera nodiflora		*Scirpus inundatus*	*(Isolepis inundata)*
Polygonaceae		*Scirpus prolifer*	*(Isolepis prolifera)*
Rumex brownii		*Scirpus nodosus*	*(Isolepis nodosa)*
Polygonum minus	*(Persicaria decipiens)*	*Schoenus brownii*	*(Schoenus apogon)*
Proteaceae		*Carex inversa*	
Xylomelum pyriforme		Gramineae	(Poaceae)
Grevillea linearis	*(Grevillea linearifolia)*	*Paspalum distichum*	
Banksia spinulosa		*Panicum sanguinale*	*(Digitaria sanguinalis)*
Banksia integrifolia		*Panicum parviflorum*	*(Digitaria parviflora)*
Thymeleae	(Thymelaeaceae)	*Panicum gracile*	*(Paspalidium gracile)*
Pimelia linifolia (sic.)		*Panicum marginata*	*(Entolasia marginata)*
Euphorbiaceae		*Panicum marginatum*	
Poranthera microphylla		var. *stricta*	*(Entolasia stricta)*
Phyllanthus ferdinandi	*(Glochidion ferdinandi)*	*Hemarthria compressa*	*(Hemarthria uncinata)*
Urticaceae	(Moraceae)	*Andropogon refractus*	*(Cymbopogon refractus)*
Ficus rubiginosa	*Imperata arundinacea*		*(Imperata cylindrica* var. *major)*
Casuarineae	(Casuarinaceae)	*Anthistiria ciliata*	*(Themeda australis)*
Casuarina glauca		*Microlaena stipoides*	
Orchideae	(Orchidaceae)	*Dichelachne crinata*	
Dendrobium linguiforme		*Dichelachne sciurea*	*(Dichelachne micrantha)*
Amaryllideae	(Hypoxidaceae)	*Deyeuxia forsteri*	*(Agrostis avenacea* var.
Hypoxis hygrometrica			*avenacea)*
Liliaceae	(Smilacaceae)	*Danthonia semiannularis*	
Smilax glyciphylla		*Cynodon dactylon*	
Dianella revoluta	(Liliaceae)	*Sporobolus diander*	
Dianella caerulea		*Eragrostis leptostachya*	
Eustrephus latifolius	(Philesiaceae)	*Eragrostis brownii*	
Commelynaceae	(Commelinaceae)	*Festuca duriuscula*	*(Festuca asperula)*
Commelina cyanea		Filices	(Pteridophyta)
Juncaceae	(Xanthorrhoeaceae)	*Pteris incisa*	*(Histiopteris incisa)*
Xerotes longifolia	*(Lomandra longifolia)*	*Asplenium flabellifolium*	

Indigenous Trees in the Gardens (1988)

The only surviving eucalypts from the natural forest in the Gardens are two Forest Red Gums, *Eucalyptus tereticornis*, on Lawn 25 east of the Maiden Pavilion, and Lawn 28. There was another old eucalypt that was probably a natural hybrid between *Eucalyptus botryoides* and *Eucalyptus resinifera* on Lawn 59 west of Bed 125 (now removed, 1992).

A number of non-eucalypt trees that are remnants of the original vegetation still occur in the Gardens.

A small grove of Swamp Oaks (*Casuarina glauca*) is to be found on Lawn 38 near the Maiden Pavilion. These would have been near the high-tide mark of the original shoreline (although Swamp Oaks grew up to the site of the State Library of New South Wales). Though not the original trees, the ones in the Gardens are suckers from an original stand and therefore biologically (genetically) the same invididuals. Further down towards the harbour (Bed 99, around a sandstone outcrop) is a grove of Cheese Trees (*Glochidion ferdinandi*), a small tree that is still common in harbourside suburbs of Sydney. This species regenerates naturally from seed in rocky places in the Gardens and the Domain.

Some of the Port Jackson Figs, *Ficus rubiginosa*, may be part of the original vegetation (e.g., plants growing on cliff edges around Beds 106 and 110, and especially those on the cliff ledges along the western side of the road going to Mrs Macquaries Chair). This is a common species in such habitats around the harbour foreshores. Seedlings seem always to appear above the ground, especially in rock walls, buildings, and the tops of palm trees. It is a modified 'strangler' fig — usually embracing rocks instead of other trees.

In *A Guide to the Botanic Gardens, Sydney*, in 1903, J. H. Maiden referred to a *Melaleuca styphelioides* or Prickly-leaved Tea Tree (west side of Bed 59): 'This is a very old tree, perhaps one of the oldest inhabitants, and very likely spontaneous; but, if not, it must have been planted at a very early period of the garden'.

Grove of Swamp-oaks in Lower Garden, 1908. These were part of the original vegetation, now reduced to three specimens, *Government Printing Office.*

Peter Hind's Checklist of Native Plants in the Domain (1988)

In 1988 Peter Hind took Maiden and Camfield's checklist of native plants in theDomain (1902) and tried to assess what remained. Virtually all of the native plants had disappeared from the Phillip and adjoining precincts of the Domain; the majority of native plants remaining being found in the Yurong and Woolloomooloo precincts, especially along the waterfront and cliff lines. It was interesting to note that there were more native ferns collected in 1988 than 1902, the ferns being the only group enriched by the disturbances.

Replanting of the eastern side of Mrs Macquaries Point with native species has continued since 1990, with emphasis on the area south of the Boy Charlton Pool.

Trees and Vines
(mostly on top towards the point)
Angophora costata
Banksia integrifolia (most probably planted)
Casuarina glauca
Clerodendron tomentosum
Dodonaea triquetra
Cupaniopsis anarcardioides (Cupania anacardioides planted in Domain, opposite Garden Island, *c.* 1847)[5]
Elaecocarpus reticulatus

Eucalyptus botryoides
Eucalyptus gummifera
Eucalyptus resinifera
Eucalyptus robusta
Eucalyptus pilularis
Eucalyptus tereticornis
Ficus rubiginosa
Glochidion ferdinandi
Kennedia rubicunda
Muellerina celastroides or eucalyptoides (mistletoe)
Pandorea pandorana
Pittosporum undulatum (seed collected, June 1988)
Rapanea variabilis
Glycine sp.

Ferns and Allies
(cliff lines and around soaks)
Asplenium flabellifolium (on Maiden's list)
Blechnum ambiguum (spore collected, June 1988)
Cheilanthes sieberi (overlooked by Maiden?)
Cyrtomium falcatum (introduced from Japan)
Davallia pyxidata
Doodia caudata
Gleichenia dicarpa
Histiopteris incisa (on Maiden's list)
Hypolepis muelleri
Nephrolepis exaltata or *cordifolia* (introduced)
Psilotum nudum
Pteridium esculentum (overlooked by Maiden?)
Pteris tremula
Pteris vittata

Herbaceous Plants
(cliff lines)
Commelina cyanea
Cotula australis
Crassula sieberiana
Cynodon dactylon
Dianella caerulea
Dianella revoluta
Dichondra repens
Cyperus mirus } (seed collected, June 1988)
Cyperus polystachyos }
Juncus sp.
Lobelia alata (seed collected, June 1988)
Lomandra longifolia
Opercularia aspera
Portulaca oleracea
Rumex brownii

Various grasses and sedges were also collected and not fully identified, but the dominant ones are introduced — Kikuyu, Buffalo and Guinea grass.

Chronological Checklist of Memorial Trees

Trees, like statues, are rarely venerated unless they are more than 100 years old. Many special trees have been planted in the Gardens over the years. Some of them have thrived, others have died, while still others may not have been properly recorded.

The following is an attempt to document as many of these special trees as possible — from a wide range of sources — up to the period of the end of the celebration of 200 years of European settlement in Australia. Some of the memorial trees listed in J. H. Maiden's article, 'History of Sydney Botanic Gardens',[6] cannot now be located, and have probably died. Some of the more recent plantings that have died are included for historical interest. The following list is adapted from Maiden and other sources, and includes all historic or interesting trees whose plantings were recorded, as well as the plantings of the last decade.

1816
Tree: *Eucalyptus robusta* (Swamp Mahogany)
Planted by: Jack Wright (and others)
Location: Lawn 26, northern side of Macquarie Wall
Additional Information: According to Maiden, this avenue was planted to define part of Mrs Macquaries Road in 1816 by a Jack Wright (and others) who had a potato stall in the markets.[7] As of 1988 these trees were still extant, making them the oldest row of street trees planted in Australia.

There were also 'some Blackbutts planted by the same hands from "the run of water" [the creek] and ran up [to, and] across the [old] fence, nearly to [Old] Government House gates'. None of these has survived.

1816
Tree: *Quercus robur* (English Oaks)
Planted by: Mr Bigg (no other details known)
Location: Along Mrs Macquaries Road from Domain entrance (near State Library of New South Wales) to St Marys entrance to the Domain
Additional Information: The brothers Charles and George Kellick (who were born in Phillip Street) obtained the history of these plantings from their father who knew

Woolloomooloo Gate, *Jaime Plaza.*

Native gums in the Domain, *Jaime Plaza.*

Clivia miniata in bloom in Palm Grove, *Jaime Plaza.*

Swamp Mahogany trees planted (*c.* 1816) near Macquarie Wall in Gardens, 1989, *Jaime Plaza.*

Bigg well, and told it to J. H. Maiden. The trees were planted by Mr Bigg on the orders of Governor Macquarie but none now survives. Apparently there was a tree left over which Mr Bigg planted in front of his house in Phillip Street. This same tree grew for many years on the path in front of the former office of the Inspector-General of Police, before it was cut down.[8]

1815/1817
Tree: *Araucaria heterophylla* (Norfolk Island Pine) — The Wishing Tree
Planted at: Mrs Macquarie's behest (see below)
Location: Intersection of the two major pathways of the Middle Garden
Additional Information: A Norfolk Island Pine, *Araucaria heterophylla*, was planted in the developing Gardens at the request of Elizabeth Macquarie, the Governor's wife, in 1815. It is reputed to have come from Norfolk Island in a tub and was first planted in the Old Government House garden, at the corner of what are now Phillip and Bridge Streets.

Norfolk Island was very much part of the consciousness of Colonial Australia, predating the settlement itself. James Cook, discovering the island on his second voyage in 1774, had noted its flax plants and pine trees, and suggested the trees may make useful ships' masts. Twenty-one days after founding the main colony, Philip Gidley King was sent 1,600 kilometres to annex Norfolk Island as a potential source of spars, plus sails and ropes from flax. Britain was the senior maritime power of the day, and wanted to stay that way. The 'pine' trees turned out to be unsuitable for spars and the flax industry failed because the local 'flax' was different from the European varieties.

It is probable that the first ships returning from Norfolk Island brought seedling pine trees back with them. François Péron, visiting Sydney in 1802 as part of the Baudin expedition, speaks of the Norfolk Island

Norfolk Island in a tub and was first planted in the old

Pine growing in the Governor's garden (old Government House site). This tree is shown in drawings and paintings of the time. One of the Eyre's 'Views of New South Wales' in 1810 shows a Norfolk Pine in front of old Government House at least thirty feet high!

James Milson, a free settler with an interest in farming, acted as Governor Macquarie's confidential land steward and superintendent of the Government farm (on the site of the present Botanic Gardens) from 1810. As part of his duties he was responsible for laying out the grounds of Government House with trees and ornamental shrubs. Supplies of seeds were received from Kew Gardens, and a number of young pines from Norfolk Island were successfully grown from seed by Milson and distributed to settlers. Therefore the first Norfolk Island Pines in the Botanic Gardens were probably planted by Milson.[9]

Dr D'Arcy Wentworth, who held several official appointments on Norfolk Island, planted Norfolk Island Pines on his property at Homebush, and on 28 June 1814 Governor Macquarie says:

> Dr Wentworth, having kindly given us a present of two large Norfolk Island pine trees from his farm at Homebush, twelve miles distant from Sydney, they were brought down this afternoon from thence in carts to the Government Domain; and on the same day the largest of the trees was planted in the Lower Park, and on the day following the smallest was planted at the intended Grass Entrance into the new Government Garden in Farm Cove.[10]

The Wishing Tree, 1892, *RBG Library*.

It is therefore apparent that a number of Norfolk Island Pines were planted in the Domain and Gardens in those early days, and had to have water carried to them in the drought of 1846.[11] Was this 'smallest' tree the Wishing Tree? That would depend where 'the intended Grass Entrance to the new Government Garden' was.

A particular Norfolk Island Pine, planted at the point of intersection of the two main pathways in the Gardens, prospered and grew to become 'part of the life of the people' — and at some stage became known as the 'Wishing Tree'. (See 'I Wish' Statue p.96). Certainly by 1846 the Norfolk Pines were said to 'arrest the attention of the stranger'.[12]

In earlier times people believed that spirits lived in trees and would not pass certain 'sacred' trees without stroking their bark to appease the tree spirits and, sometimes, to make a wish. Thus the idea of 'wishing trees' developed, and the expression 'touch wood'. The custom grew around the Sydney Wishing Tree that children and lovers should walk around it six times — three times forward, three backward — and then make their secret wishes.

In November 1905 the Wishing Tree was measured by the Gardener, H. Malthouse, who found the height to be 99 feet 3 inches (30.25m). It was feared that its maximum altitude had been attained; a few years before being slightly over 100 feet.[13]

There has been considerable confusion about the Wishing Tree and many people, struggling for a footnote in history, have laid some claim to having planted it. It may have been one of James Milson's trees, or one of the trees given to the Governor by Dr Wentworth, the one from Old Government House Garden — or indeed another seedling from Norfolk Island. According to Walter Froggatt, quoting J. H. Maiden,[14] the Wishing Tree was first planted in front of Old Government House in Bridge Street. Then in 1816, at Mrs Macquarie's request, it was removed to the Botanic Gardens and planted near the stream (Botanic Gardens Creek).

Early in 1872, there was a picture of the Wishing Tree in the *Illustrated Sydney News* with a caption saying it was not known when it was planted. This brought a letter to the editor from John Richardson, who wrote:

> My attention having been drawn to an article in your paper relative to the large Norfolk Island Pine in the Botanic Gardens, in which you state that it is not known when the said tree was planted. I beg to inform you that I planted it in the year 1818, at which time I was employed in the Gardens. It had been previously planted in the Lower Garden, near the watercourse, and had been washed out during some heavy rains which had occurred about that time. It was lying half buried in the sand for almost six months, when it was removed and planted by me in its present situation in the year above stated. Mrs Macquarie was present on the occasion and handed me the tree to plant in the ground. It was then about two feet six inches in height.[15]

This would rule out the Wishing Tree as having been the one transplanted from in front of Old Government House.

Some years later Major-General J. M. Antill, in newspaper correspondence described the planting of two Norfolk Island Pines in the Botanic Gardens by his father and Mrs Macquarie in 1816. He said one of these was called the 'Antill Pine' and later became the Wishing Tree.[16]

Still others laid claim to planting the Wishing Tree — including Charles Fraser, Superintendent of the Gardens; Ned Shakely, a convict later employed at John Baptist's Surry Hills nursery; John Higgerson, ranger of National Park (died 1905); and Mrs Macquarie herself. Despairing of the conflicting legends, Maiden, who considered the Wishing Tree to be the 'most historically interesting plant in our beautiful Garden' boldly engaged in a little playful historical fantasy

> I propose to adjust these claims in the following manner: Ned Shakely dug the hole, Johnny Higgerson handed him the spade and helped him generally, Mr Fraser turned the plant out of the pot to see that it was all right, and Major Antill planted it with due ceremony. Then Mr Fraser trod the earth about it, staked it, watered it and tendered it during its early days.[17]

And where was Mrs Macquarie when all this was going on?

Having attended numerous tree plantings (and replantings!) I would now like to propose the following construct, including John Richardson's letter and other sources. In 1816 Major Henry Antill, an American Loyalist officer and Macquarie's aide-de-camp, was in charge of the operation to plant a couple of Norfolk Island Pine trees in the developing Gardens, assisted by two privates from the 73rd Highlanders. Mrs Macquarie probably planted the trees, possibly in the presence of Charles Fraser (who had arrived on the *Guildford* on 8 April 1816). The holes were dug by a convict called Ned Shakely. Johnny Higgerson may well have helped. Two years later one of these trees was uprooted by a storm (because it had been planted near the creek) and lay half-buried in sand for six months before being replaced by Mrs Macquarie, with help from a gardener's boy, John Richardson. This is the tree that grew to become the much-loved Wishing Tree.

On 25 August 1989 a young man called Peter Euston called in at the Visitor Centre inquiring about the Wishing Tree. He told me his grandmother, Mrs Essie Euston of Newcastle, who was born in 1903, was the great granddaughter of John Richardson. According to Peter Euston the family legend was that John Richardson was a horticulturalist who had come from England to the Sydney Gardens 'when there was virtually nothing there' and had helped the Governor's wife plant the Wishing Tree. John Richardson later settled in Newcastle. Peter Euston said his grandmother had a piece of the original tree — presumably one of the souvenirs made from the wood when it was cut down.

It is not known precisely when the tree was called the Wishing Tree, but Ethel Turner, writing in 1895 of her childhood days, recalled:

> There was not time to 'do' the Gardens thoroughly, so they only walked rapidly up some of the paths, paused for a moment to look at the blue harbour beyond the low sea wall, and then walked three times solemnly and backwards around the wishing tree near the entrance gates.[18]

The Wishing Tree was finally removed in 1945, when old age and decay threatened to topple it, and souvenirs were made from its wood and sold to help disabled soldiers returning from the war.

Ten years earlier, when it became apparent that the old tree was dying, Sydney children contributed to buy a replacement tree, which was planted near Bed 40, just north of the Macquarie Wall on 27 July 1935 by Lady Hore-Ruthven, wife of the State Governor. The replacement tree, however, has never thrived, and has never engendered the same affection as the original in this more cynical age. Robert Henry Anderson, a former Director of the Gardens, summed this up nicely in 1964 when he said, 'The new Wishing Tree had made little impact on the public and is not generally accepted as having any special powers'.

The Wishing Tree

(Ode to Mrs Macquarie's Wishing Tree, planted Royal Botanic Gardens, Sydney)

She had a seedling planted where
the pathways crossed, close to a martyr's
unmarked grave, a Norfolk Pine
that grew into a tall ship's mast;

where young girls tripped thrice,
forwards, back, and lipped priorities
of love and rank; near beds
that bear the mark of colony.

The tree outlived its dignity,
put down when he was just a boy;
tall poppies wave their secret joy,
that numbs the mind to servitude.

A phantom tree, antennae clipped,
still pulls the static of the past,
a plaque, a statue, forties kitsch;
arthritic earth: I wish, I wish.[19]

1828

The following trees were probably collected by Charles Fraser (and Allan Cunningham) on the banks of the Brisbane River in 1828:

Tree: *Araucaria cunninghamii* (Hoop Pine)
Location: Bed 23
Tree: *Flindersia xanthoxyla* (Yellow-wood)
Location: Bed 25
Tree: *Waterhousia floribunda* (Weeping Lilly Pilly)
Location: Bed 26
Tree: *Cryptocarya obovata* (Pepperberry)
Location: Lawn 6
Tree: *Aphananthe philippinensis* (Native Elm)
Location: Bed 26
and possibly
Tree: *Toona ciliata* (formerly *australis*)
Location: Bed 9
Tree: *Melaleuca quinquenervia* (Broad-leaved Paper Bark)
Location: Bed 10

The last two trees occurred locally and may well have been planted earlier.

Additional Information: At least 12 trees were brought back from Queensland in 1828 by Charles Fraser, who accompanied John Oxley and Allan Cunningham on their exploring trip to the Brisbane and Logan rivers, at which time there was a 'dense forest on the southern bank' of the Brisbane River. On 16 and 17 July 1828, Fraser obtained living plants in the vicinity of (Brisbane) town, and on 22 July he was 'employed in procuring plants on the river banks until eight o'clock.

Returning from his trip to the south west on 11 August, Fraser expressed 'A strong desire to return by the first vessel to Port Jackson, carrying thither the largest possible collection of living plants from the banks of the Brisbane River'. Of those, as many as seven have possibly survived. They are 'magnificent trees which have been growing steadily during the greater part of the period of Australian [settlement, and] combine the dignity of age with this most interesting historical association'.[20]

Charles Moore, in his Report to the Legislative Assembly, 1870–71, reported:

> On the western side [of the Old (Middle) Garden] . . . are fine examples of some of the trees which constitute the brush or jungle forests of our Coast Districts, collected and planted here by my predecessors, Cunningham & Fraser. Among these are the first discovered plants of some now widely distributed and well known trees — viz. Silky Oak, Moreton Bay (Hoop) Pine, *Araucaria cunninghamii*, and others.[21]

The Silky Oak is no longer there. It is a great pity he did not list the others.

Following correspondence with Dr T. D. Macfarlane, Australian Botanical Liaison Officer at Kew, London, in 1990, Joy Thompson (Sydney Gardens) assisted by consulting the archives and had discussions with Dr L. A. S. Johnson, Director Emeritus, (who was familiar with the plantings referred to and also visiting Kew at the time). There was no Fraser correspondence, but the Cunningham correspondence[22] indicates that he was to collect for Sydney as well as Kew from early 1828.

Collections from this time were sent to Kew but few of these specimens now appear to be in the herbarium. Apart from the *Flindersia* (listed as collected Brisbane R., Moreton Bay, A. Cunningham, July 1828) the other Cunningham collections were from conflicting dates and places. No reference to *Toona ciliata* (formerly *australis*) was found, although *Melaleuca quinquenervia* was collected at the Logan River by Fraser in 1829.

1830
Tree: *Salix babylonica* (Willow)
Planted by: Charles Fraser
Location: Near where Gardens Restaurant now stands
Additional Information: Willows were probably established in the Botanic Gardens before 1803, for they were recorded in Governor King's list of plants in the colony for that year.[23] In 1830 there was a significant mention of the Weeping Willow in the Gardens:

> Mr Fraser, the Colonial Botanist, has a cutting from the willow which hangs over the tomb of Napoleon at St Helena, and is about to plant it on one of those miniature islands constructed by him in the pond in the New Botanic Gardens which he is laying out in Farm Cove.[24]

The miniature islands referred to were in the Lower Garden, near where the Gardens Restaurant now stands. On one of these islands the monument to Allan Cunningham was erected in 1844. It must be remembered that before the Sea Wall was built the high tide in Farm Cove extended up to this point. The Cunningham memorial is now situated in one of the ponds near the Restaurant.

Charles Fraser was not to live much longer. He collapsed at Emu Plains and died at Parramatta on 22 December 1830, aged forty-three.[25]

The willows growing on the original island in 1931 were grown from cuttings from this original.[26] J. H. Maiden had earlier indicated that 'all the weeping willows in the vicinity of this pond and most of them in other parts of the Garden are descendants of Napoleon's willow'.[27] It is not known if the willow growing in Lawn 40 (and an earlier one that grew there, near the largest pond in the Lower Garden) is a cutting from these earlier trees.

1840
Plant: *Livistona chinensis*
Location: Two specimens at southern ends of Beds 18 and 19
Additional Information: These two plants were transported in tubs in 1840 from the Isle of Bourbon (Reunion Island, Indian Ocean) and planted in their present positions in December of that year. Their growth rate has been incredibly slow.

Fig Tree Avenue (planted 1847) in the Domain, *c.* 1880, *Government Printing Office.*

1847
Tree: 'Fig Tree Avenue' *Ficus macrophylla* (Moreton Bay Figs)
Planted: During the administration of Charles Moore
Location: Domain, from region where the Shakespeare Group Statue now stands to what was Main Entrance of Gardens, near Restaurant on the Park (opposite Art Gallery of New South Wales).
Additional Information: The Moreton Bay Figs (planted in 1847) grew into the celebrated 'Fig Tree Avenue', destroyed during the construction of the Cahill Expressway, *c.* 1960. The truncated remains of some of them are now marooned on a traffic island in the Cahill Expressway; and others were planted *c.* 1986.[28]

1853
Tree: *Agathis robusta* (Queensland Kauri) and *Agathis moorei* (New Caledonia)
Planted: During the administration of Charles Moore
Location: Bed 57

Current site of tree planted by Duke of Edinburgh in 1868, *RBG Library*.

Paper-bark planted by HRH Prince George in 1881, *RBG Library*.

Additional Information: The Queensland Kauri, one of the tallest in the Gardens, was planted in 1853.[29] It is believed to be the same tree described and illustrated in 1857 in an anonymous article in a Sydney-based literary and critical journal, *The Month*. The plant was identified by the genus *Dammara*, and subsequently described by Baron von Mueller.

1856
Planting: Spring Walk
Planted: During the administration of Charles Moore
Location: Beds south of Macquarie Wall
Additional Information: A number of Rhododendrons were planted on the southern side of the wall on 10 June 1856. In 1871, Charles Moore also reports, 'a large number of . . . Rhododendrons was planted . . . nearly all . . . have perished . . . Azaleas, however . . . succeed admirably here, and flower every year most profusely'.[30] The area later became known as the Spring Walk, as indeed the path on the southern side of the wall is so-named today.

In 1888 Charles Davis reported that there were some fine specimens of elkhorn ferns growing on the remains of the old wall, and that they had been there for about thirty years.[31]

1862
Planting: Palm Grove
Planted: During the administration of Charles Moore
Location: Beds 23, 24, 25, 26, 27, 28, 57, 58, 59
Additional Information: The first plantation of more hardy palms was made in 1862. A second plantation of more tender palms was made in 1864.[32]

1868
Tree: *Chrysophyllum (Martiusiella) imperialis*
Planted by: Prince Alfred, Duke of Edinburgh
Location: Southern end of Bed 18 — originally planted in 'Band Lawn' (Lawn 24)
Additional Information: This rarely grown Brazilian rainforest tree from the diminishing forests near Rio de Janiero was planted by Prince Alfred, Duke of Edinburgh (one of Queen Victoria's sons, when he was visiting Sydney in the naval ship, *Galatea*) in 1868. The Prince was later to be the victim of an assassination attempt at Clontarf.

1881
Tree: *Melaleuca quinquenervia* (Paper Bark)
Location: Lawn 24
Planted by: HRH Prince George (later King George

Cabbage Palm planted by the Duke of Clarence in 1881, photographed in 1901, *RBG Library*.

New Caledonian Pine planted in 1901 by the Duchess of York (later Queen Mary) and lost in a severe storm in 1990, *RBG Library*.

V) or HRH Prince Albert (later Duke of Clarence)

Additional Information: On 22 July 1881, two royal princes, Albert Victor Christian Edward (*c.* 1864–1892) and George Frederick Ernest Albert (1865–1936) — sons of the Prince of Wales (later King Edward VII) and Queen Alexandra — planted two trees in Lawn 24 (Band Lawn) at the Sydney Botanic Gardens.

This event, as later published by their tutor, the Reverend John Neale Dalton, in a book based on excerpts from their private diaries and letters, was recorded as follows: 'Planted a couple of trees in a Botanical Garden, close to the spot where, 2 April, 1868, The Duke of Edinburgh did the same'.[33] There is no doubt that this relates to the two 'Royal' trees on Lawn 24; the *Chrysophyllum imperiale* (synonym *Martiusella imperialis*) planted earlier by Prince Alfred, Duke of Edinburgh in 1868 in the 'Band Lawn', was later moved to its current site in the southern end of Bed 18. The two trees planted by the royal princes were *Melaleuca quinquenervia* (Paper Bark) and *Livistona australis* (Cabbage Tree Palm) but it is impossible for us to know which prince planted which tree as they did 'everything together'.[34]

The princes were 'in Sydney from 15 July to 10 August 1881, as part of an astonishing three-year tour which began when they were aged 15 and 14 respectively'.[35] There was some considerable confusion as to their names as published in the papers at the time, and the repeated quoting of these earlier accounts has only compounded the problem.[36] The elder of the two princes, Albert Victor Christian Edward (later to become the ill-fated Duke of Clarence), was known in the Royal Family as 'Eddy'. The younger prince, George Frederick Ernest Albert replaced his brother (who died prematurely, in 1892) as heir, to become King George V in 1910.[37] Prince George was much brighter than his older brother Albert.

> Even the most tactful of Royal biographers refer to [Albert's] lethargy, listlessness, and complete inability to study. Doubt has been expressed as to whether he actually learnt to read. His dissolute life was a cause of scandal . . . and a well-known modern writer [has] suggested that the 'little prince' became 'Jack the Ripper'. He died, officially, of 'pneumonia', on 14 January, 1892, aged 28.[38]

At the end of the tour Dalton produced his 'extraordinary' book, *The Cruise of HMS Bacchante*, . . . in which he had the young princes quoting long passages from the Vulgate in the original Latin . . . as well as being able to quote from Theocritus, Browning, Shakespeare, Byron and Tennyson.[39]

In 1981, leading up to the centenary of the planting of their planting, a birthday party was arranged for these trees. The *Melaleuca quinquenervia*, being the more robust, was arbitrarily deemed to have been planted by the later King. Pupils from the Neutral Bay Primary School Environment Club, who had been one of the first school groups to contribute to the Pyramid Glasshouse Appeal, were invited to come to the party and help eat the birthday cake, which had been especially prepared for the event by the Food School, East Sydney Technical College.

1901

Tree: *Araucaria columnaris* (New Caledonian Pine)
Planted by: Princess Mary, Duchess of York (later Queen Mary)
Location: Lawn 53
Additional Information: Planted as part of the celebrations to commemorate the Federation year. This tree was struck by lightning in February 1990 and snapped off above the ground during a severe storm on 3 August 1990.

During this visit of the Duke and Duchess of Cornwall and York, the Duke planted a *Stenocarpus sinuatus* (Firewheel tree) near the New Caledonian Pine which no longer survives. The Duke also planted a Port Jackson Fig at Garden Island (declared to be a portion of the Domain in 1811) and the Governor-General, Earl Hopetoun, planted a Hoop Pine on behalf of Her Royal Highness. Her Royal Highness also planted a 'Blue Mountain Cypress Pine' (*Callitris* sp.) at Admiralty House, between two additional trees planted by the two princes on their visit to Australia on the *Bacchante* in 1881.[40]

1904

Tree: *Casuarina cunninghamiana*
Planted by: Governor-General Lord Northcote and Lady Northcote
Location: Near Inner Domain Gate, opposite Bed 133[41]

1916 — Centenary Trees

Tree: *Ceratonia siliqua* (Carob tree)
Planted: 'for Serbia'
Location: Lawn 53, south west of Bed 125 — removed 15 October 1987 — cuttings taken
Tree: *Populus deltoides*
Planted: 'for Canada'
Location: Lawn 58
Tree: *Eucalyptus maidenii* (Maiden's Blue Gum)
Planted: 'for New South Wales'
Location: Lawn 58
Tree: *Agathis robusta* (Queensland Kauri)
Planted: 'for New Zealand?' (may have been a misidentification of seedling!)
Location: Lawn 58
Tree: *Quercus robur* (English Oak)
Planted: 'for England'
Location: Lawn 58
Tree: *Pinus pinaster*
Planted: 'for Portugal' (may be a replant — A. N. Rodd)
Location: Lawn 58
Tree: *Ficus rubiginosa* (Port Jackson Fig)
Planted: 'for Sydney'
Location: Lawn 59
Tree: *Harpephyllum caffrum* (Kaffir Plum)
Planted: 'for South Africa'
Location: Lawn 59
Additional Information: A number of trees were planted on 13 June 1916, to commemorate the Centenary of the Botanic Gardens, in the region of the Parade Grounds (Lawn 58) adjacent to Government House. Various trees were planted by representatives of Allied Nations (in what was the middle of World War 1). A contemporary map of the time has enabled us to identify these trees with a high degree of certainty.

1934

Tree: *Morus nigra* — Shakespeare's Mulberry Tree
Planted by: Shakespeare Society of New South Wales
Location: Bed 84
Additional Information: See p.63, 1987 — replanting of cutting from Shakespeare's Mulberry Tree.

1935

Tree: *Araucaria heterophylla* (Norfolk Island Pine) — replacement Wishing Tree
Planted by: Lady Hore-Ruthven, wife of the State Governor
Location: Circular bed north of Macquarie Wall and east of Bed 41

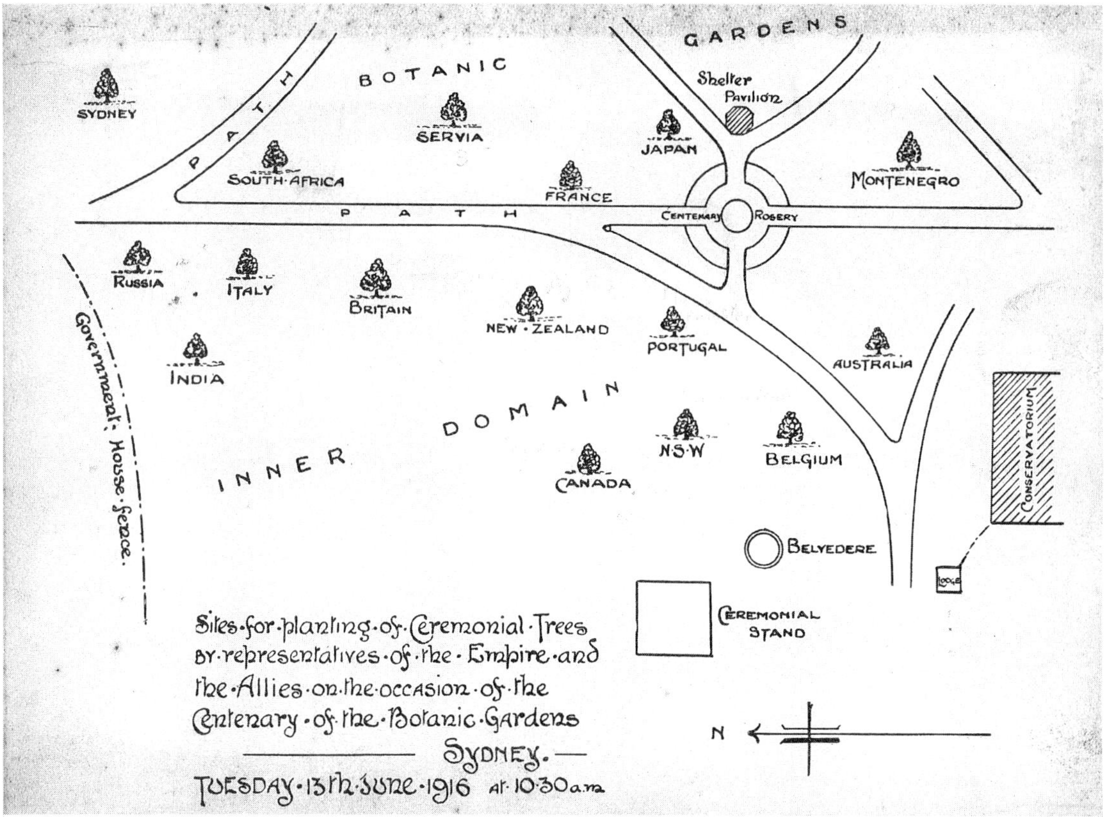

Sites for Centennial Plantings in the Gardens, 1916, *RBG Library*.

Additional Information: When it became apparent that the Wishing Tree was dying, Sydney school children contributed to buy a replacement tree. However, it has never engendered the same affection as the original Wishing Tree.

1935
Tree: *Quercus robur* (English Oak) — Mons Oak
Planted by: Squires of Mons
Location: Lawn 57
Additional Information: The Mons Oak, planted in the Royal Botanic Gardens Sydney on 25 August 1935, by the Squires (sons) of Mons (21 years after the battle fought by their fathers), was grown in soil taken from the grave of an 'unknown soldier' who fell during the British retreat from the Belgian town of Mons in 1914.

According to an article by Bill Mellor, written on the fiftieth anniversary of the planting of the tree, the British were 'hopelessly outnumbered and outgunned' as they fought a 'desperate rearguard action through the Ardennes' ... but instead of delivering the *coup de grâce*, the Germans abruptly halted their advance.[42] After the war the *Church Times* carried an interview with a German soldier who had participated in the Battle of Mons. The German claimed that he and his colleagues had wanted to go forward and mop up the British, but had found themselves inexplicably 'powerless to proceed ... our horses turned sharply round and fled like the wind'.

In the 'half-light of a false dawn' a platoon of Coldstream Guards saw a 'tall, female figure in a flowing white robe who beckoned them to follow her ... across an open field ... to a sunken road they did not remember crossing ... back to the safety of the main British forces'. The starving, exhausted and shell-shocked men may well have hallucinated that morning on the battlefield; survivors claim they saw an angel and other 'apparitions in the sky'. A story of 'divine intervention ... spread like wildfire through the British lines. But despite many a search, no-one ever found that road'.

In a subsequent article in the *Sydney Morning Herald*,

Alan Gill elaborated still further on this mystery.[43]

Sceptics claimed it to be a hoax dreamed up by a Fleet Street writer Arthur Machin, who claimed in a book, *The Bowmen and Other Legends of the War*, to have invented the angel of Mons. Another author, Harold Begdie, sided with the angels, in claiming the story had circulated well before Machin had heard about it.

Then, in April 1930, a former German Intelligence officer, Colonel Friedrick Herzenwirth, claimed that the 'Angel of Mons' had been produced by magic lantern slides thrown upon 'screens of foggy cloudbanks by projectors with high-powered Zeiss lenses mounted on German aeroplanes which hovered above the British lines'. The story was dismissed as implausible by technical experts, but the German stuck to his explanation. He said the vision, intended to frighten the British, had misfired because the Guardsmen assumed the angel was their protectress.

Colonel Herzenwirth said the Germans had achieved better luck with their cloud magic lantern shows on the Russian front in 1915. The Virgin was shown with uplifted hand, as if motioning to stop the 'murderous' Russian night attacks. 'We know from prisoners we took,' said the colonel, 'that in some cases companies actually killed their officers and flung their rifles away, shouting that they would not be guilty of firing upon an army over which the Mother of God hovered in protection'.

However, with the French, in Picardy, and the Champagne region, the Germans made another miscalculation, for 'instead of taking the figure of a woman that we threw upon the clouds one night as that of the Virgin or a saint protecting [the German] army, the French promptly recognised Joan of Arc'.

Letters in the Imperial War Museum, London, and the Australian War Memorial suggest that the angel appeared in different locations, though at approximately the same time. There is disagreement about the angel's gender. None of the letters has any reference to 'beckoning', but several claim that there was not one 'angel' but three. The most interesting statement — couched in terms which may support the magic-lantern theory — is an account by a Scots Guards corporal, dictated from his hospital bed. This witness saw 'three figures . . . midway between the earth and the sky . . . over the German lines and facing towards the British': 'They kept on growing brighter and brighter. The centre figure was taller than the other two and had shining wings which seemed to protect the lesser figures . . . under the feet of the three figures was a bright star. When the figures disappeared the star remained.'

Alan Gill concluded his article with a reference to a story in the 1920s, in the *Observer*, about a 'bright young thing' who challenged the veracity of the Angel of Mons at a party. A colonel looked up and said, 'Young lady, the thing happened. You need not be incredulous. I saw it myself.'

The day before the fiftieth anniversary of the planting, 24 August 1985, while reading the Bible, I inadvertently came across the reference to Byron's poem, 'The Destruction of Sennacherib', in II Kings xix.35. These two images then overlapped and lay dormant in my mind for almost two years. Then on Bastille Day 1987, while passing the Mons Oak, I was compelled to read its plaque. The following poem, which honours valour, atonement, and comradeship (and not that 'sordid trade war') fell into my head on that day, to be polished on the next:

Mons Oak

The battalion at Mons was outgunned and
 outflanked,
back to the Ardennes, in the top end of France,
and the battle was lost, and the battle was
 won —
when a luminous host stole a march on the
 Hun.

For the Angel of Mons held her sword to the
 sky,
and rode a white horse into mayhem and wire;
and led the fazed men to a road never crossed,
where a Coldstream platoon had been hopelessly
 lost.

She appeared in a false-dawn of cordite and gore,
and delivered them back to the main British
 force;
on a road to redemption through valour and
 mud —
conned by their own faith in a temple of blood.

The word spread like fire along the front line —
a Division had bridged the apparent divine,
outflanked by a girl in a cirrus-white robe;
and try as they might, they could not find that
 road.

They carried back soil in an old ammo tin,
from the grave of a comrade, that valour should
win;
and planted an oak in that sanctified clay —
the Companions of Mons saw an angel that day.

As oak leaves in Autumn all memories fade,
to garnish the temples of Empire and Trade;
as spoils to the victor, the vanquished, the
gun —
that God not bless bullets of Briton or Hun.[44]

1936
Tree: *Populus* × *canadensis* cv. Aurea
Planted by: Dame Mary Gilmore
Location: Lawn 2
Additional Information: The great Australian writer Dame Mary Gilmore planted a poplar close to the Kendall Memorial seat in the so-called Poet's Corner in 1936. It was said to be a cutting from a tree planted at 'Merryvale', Roslyn, New South Wales, to commemorate her birth on 16 August 1865. This cutting had originally been obtained from the Sydney Gardens and to quote Dame Mary: 'So this young shoot has come back to the Gardens where its first Australian ancestors had root and is third in descent from one of the first poplars to come to this country'.[45]

Poplars grow rather quickly and it is assumed that the rather large poplar in Lawn 2 is indeed the one planted by Dame Mary Gilmore.

1947
Tree: *Grevillea robusta* (Silky Oak)
Planted: In memory of Robert G. Stubbs
Location: Lawn 8
Additional Information: Robert Stubbs, a young member of staff, tended the Rose Garden near the Old Herbarium building. He joined the 2/12 Field Ambulance and was killed when the hospital ship *Centaur* was sunk by enemy action, 14 May 1943. This tree to his memory was planted by his mother on behalf of his friends at work in September 1947.[46] Apparently it died in the late 1970s and the plaque was removed.

Following a letter from George Chippendale (a colleague of Robert Stubbs who played cricket with him in the Botanic Gardens Cricket Team at Moore Park on Saturday mornings) to Professor Chambers (4 August 1989) the plaque was located. A replacement tree was planted on 15 December 1989, in the northern end of Lawn 27, by his brother John, and his sister Betty King.

Botanic Gardens Cricket Club, 1938/39. Robert (Bobby) Stubbs is second from left in front row. Information provided by George Chippendale, far right in back row, *RBG Library.*

1954
Tree: *Melaleuca leucadendra*
Planted by: Her Majesty Queen Elizabeth II
Location: Government House Garden
Tree: *Eucalyptus nicholii*
Planted by: His Royal Highness the Duke of Edinburgh
Location: Government House Garden
Additional Information: These trees were planted to the west of the drive to Government House on 8 February 1954, during the first visit to Australia by a reigning monarch. Two additional trees, a Chinese Elm (*Ulmus parvifolia*) and Cape Chestnut (*Calodendrum capense*), were planted in pots to be later placed in permanent positions in the Gardens by the Governor, Sir John Northcott. Unfortunately these two trees were lost when the Cahill Expressway was built.[47]

1970
Tree: *Casuarina cunninghamiana* (River Oak)
Planted by: HRH The Prince of Wales and HRH Princess Anne
Location: Government House Garden
Additional Information: On 2 April 1970 HRH The Prince of Wales and HRH Princess Anne each planted a *Casuarina cunninghamiana* to the west of Government House.

1975
Tree: *Arbutus clandulosa*
Planted by: HRH The Princess Margaret
Location: Government House Garden

Additional Information: This tree, planted on 31 October 1975, has since died.

1978
Tree: *Ulmus minor*
Planted by: HRH Princess Alexandra, Mrs Angus Ogilvy
Location: Government House Grounds
Additional Information: Planted 27 September 1978, west of main drive.

1980
Tree: *Angophora costata*
Planted by: Rostrum Clubs of New South Wales during their Jubilee Celebrations
Location: Phillip Precinct of Domain, west of Restaurant on the Park
Additional Information: Planted 20 July 1980 — dedicated by Rostrum, 1 May 1988.

1980
Tree: *Prunus serrulata* cv. Shiratae and cv. Fugenzo (Flowering Cherry)
Planted by: Japanese Consul General, Mr M. Mizoguchi
Location: Bed 60
Additional Information: Planted 5 September 1980 — two flowering cherries.

1980
Tree: *Abarema muellerana*
Planted by: Sir Roden Cutler, Governor of New South Wales
Location: Government House Garden
Additional Information: Planted 23 December 1980, west of tennis court.
Tree: *Cerbera floribunda*
Planted by: Lady Cutler
Location: Government House Garden
Additional Information: Planted 23 December 1980 in border plantings north-east of tennis court.

1981
Tree: *Acmena brachyandra* (Red Apple)
Planted for: Womens Pioneer Society of Australia
Location: Lawn 11
Additional Information: Planted 29 March 1981 to commemorate the first Heritage Week held in NSW.

1981
Tree: *Toona ciliata* (formerly *australis*)
Planted by: HRH The Prince of Wales
Location: Lawn 24
Additional Information: Planted 15 April 1981 to commemorate the International Year of Disabled Persons, and the centenary of the other two trees planted on this lawn.

1982
Tree: *Toona ciliata* (formerly *australis*)
Planted by: Vincent Serventy (naturalist)
Location: In Domain near Sir John Robertson statue
Additional Information: Planted 19 March 1982, for World Forestry Day, leading up to the Year of the Tree.

1982
Garden: Solander Garden
Dedicated by: His Majesty King Carl XVI Gustaf of Sweden
Location: Bed 13
Additional Information: Dedicated on 31 March 1982, in honour of Daniel Solander, Swedish botanist, and featuring species collected in eastern Australia in 1770. A bust of Solander (see p.104) was unveiled at the same time.

1982
Tree: *Atalaya multiflora* (Whitewood)
Planted by: The Hon. Malcolm Fraser, Prime Minister of Australia
Location: Lawn 19, near Bed 53
Additional Information: Planted 27 September 1982 as part of the 'Greening of Australia' Campaign. *Atalya multiflora*, the Broad-leaved Whitewood, is a rare tree on the verge of extinction in the wild.

1982
Tree: *Melaleuca quinquenervia*
Planted by: The Minister for Forestry
Location: On eastern slope of Domain up from car park
Additional Information: A group of paper-barks was planted in the Crescent Precinct of the Domain.

1983
Tree: *Casuarina cunninghamiana* (River Oak)
Planted by: HRH The Prince of Wales

Location: Government House Garden
Tree: *Casuarina cunninghamiana* (River Oak)
Planted by: HRH The Princess of Wales
Location: Government House Garden
Additional Information: Both these trees were planted on 3 March 1983, west of Government House.

1983
Tree: *Ficus macrophylla* (Moreton Bay Fig)
Dedicated as: The Children's Tree
Location: Lawn 44
Additional Information: The large Moreton Bay Fig on Lawn 44, which is well over 100 years old, was dedicated as the Children's Tree in ceremonies on 10 and 11 April 1983, organised by Lucy Bucknell.

1983
Tree: *Fagraea berteriana*
Planted by: Kate Fitzpatrick
Location: Bed 63 (this tree was not there in 1988)
Tree: *Gilbeea adenopetala* (Pink Aldes)
Planted by: Sir Rupert Myers
Location: Bed 63
Tree: *Hernandia bivalvis* (Grease Nut)
Planted by: Allan Seale
Location: Bed 63
Additional Information: These three rainforest trees were planted on 19 April 1983, at a media launch for the Friends of the Royal Botanic Gardens, as part of the proposed rainforest planting in this area.

1983
Tree: *Carpinus betulus* (European Hornbeam)
Planted by: The Hon. Neville Wran, Premier of New South Wales
Dedicated to: Captain Arthur Phillip, RN
Location: Lawn 11
Additional Information: The European hornbeam is a common hedge plant from Europe to Iran, and its hardy wood is used in making piano mechanism parts. The plaque, unveilved on 3 October 1983, reads: 'Dedicated to [the] Memory of Captain Arthur Phillip, RN 1st Governor of New South Wales (whose father came from Germany). Presented by the Committee celebrating the centenary of Australian/German institutions in Australia.'

1984
Tree: *Eucalyptus umbellata* (Lockyer's Gum)
Planted by: Ken Rendall, Foreman, Royal Botanic Gardens
Location: Lawn 29
Additional Information: Planted on 4 January 1984, to mark Ken Rendall's retirement from the Gardens after 45 years and 6 months with the New South Wales Public Service (this tree was not there in 1988).

1984
Tree: *Eucalyptus parvifolia* (Boys Brigade Tree)
Location: Lawn 29
Additional Information: This tree was not there in 1988.

1984
Tree: *Synoum glandulosum* (Scentless Rosewood)
Planted by: The Hon. Neville Wran, Premier of New South Wales
Location: Bed 13
Tree: *Callistemon viminalis* (Red Bottlebrush)
Planted by: Jill Wran
Location: Bed 13
Additional Information: Two trees planted on 22 March 1984 by Neville and Jill Wran for the Sydney launch of the Plant Campaign of the World Wildlife Fund.

1985
Tree: Variety of Australian plants
Planted by: Ken Douglas, N.S.W. State Manager of Telecom, to commemmorate Telecom's 10th anniversary
Location: Mrs Macquaries Point
Additional Information: A variety of plants, representative of the original vegetation of the point, was planted on 31 July 1985 (Arbor Day).

1985
Tree: *Eucalyptus coolabah* (Coolibah Tree)
Planted by: Rolf Harris and The Hon. Laurie Brereton, Minister for Public Works, Parks and Roads
Location: Northern end Bed 121, beside Twin Ponds (billabong)
Additional Information: A Coolibah Tree, *Eucalyptus coolabah*, was planted near the Twin Ponds (billabong) on 3 April for the media launch of Heritage Week, 1985.

In his book *Great Australian Legends*, Frank Hardy suggests that it wasn't a swagman who jumped into the billabong, but that Banjo Paterson had modelled his poem on the Bavarian anarchist, Samuel Hoffmeister. In the 1890s Hoffmeister had led a group of striking shearers who had burnt down the shearing shed on Dagworth Station, in western Queensland. When McPherson, the squatter, and three troopers rode into the shearers' camp at a billabong, Hoffmeister shot himself.

Banjo Paterson stayed at Dagworth Station a year later and wrote the poem and had it set to music. If Hardy's suggestion is correct, something was lost in translation. Paterson turned Hoffmeister into a sheep-stealer and had him drown himself. Perhaps there was really mutton in the billy?

Waltzing Matilda became a popular song with Australian soldiers in the First World War — and a de facto national anthem — although I suspect such sentiments have subsequently waned. However, if, in the midst of our urban sprawl, this tree should grow to become a symbol of yearning for a nostalgic bush culture and a perceived lost innocence and frontier mentality, then let me set the record straight.

Because of frequent inquiries from tourists I requested and organised that such a tree be planted in the Gardens. The seeds were obtained by Peter Hind and plants were grown in the Depot. The seedling was planted in tandem by a performer (Rolf Harris) and a politician (Laurie Brereton) to the refrains of Waltzing Matilda sung by the Forest High School choir; Rob Grewcoe, gardener, dug the hole. In our bicentennial year the Coolibah Tree was still there, if somewhat scraggy, having survived at least one attempted decapitation by a school boy using its name tag as a tomahawk.

1985

Plant: *Cabernet sauvignon* (grape vines)
Planted by: Hon. N. K. Wran, Premier of New South Wales
Location: Beside Glasshouse 10
Additional Information: Planted on 15 August 1985, near the site of the Busby collection of vines (1833), to represent the six wine-growing Australian States. The site of the first vineyard in Australia was not in fact in these Gardens, but beside Governor Phillip's tent at Sydney Cove.[48]

1986

Tree: *Backhousia citriodora* (Sweet Verbena)
Planted by: Her Majesty Queen Elizabeth II
Location: Government House Grounds
Additional Information: Planted on 5 March 1986, on the lawn north-east of Government House.

1986

Tree: *Backhousia citriodora* (Sweet Verbena)
Planted by: HRH The Duke of Edinburgh
Location: Government House
Additional Information: Planted 27 May 1986 (twin of above) south-east of Government House.
Tree: *Ulmus parvifolia*
Planted by: HRH The Duke of Kent
Location: Government House Grounds
Additional Information: Planted on 27 May 1986, west of main drive.

1986

Planting: Myrtales Bed
Planted by: Children of Plunkett Street Public School
Location: Extension of Bed 121 onto what was Lawn 45
Additional Information: Planted on Arbor Day, 28 July 1986.

1987

Tree: *Pinus pinea*
Planted by: Queen Margrethe II of Denmark
Location: Government House Grounds
Additional Information: Planted 5 February 1987, west of tennis court.

1987

Tree: *Morus nigra* (Black or Common Mulberry) — Shakespeare's Mulberry Tree
Planted by: Ian McLellan, Foreman
Location: Lawn 17
Additional Information: The Black Mulberry (*Morus nigra*), planted in Lawn 17 in 1987, was propagated from a hardwood cutting (*c.* 1982) by Ian McLellan, Foreman, at the Royal Botanic Gardens. The cutting was taken from a plant in Bed 84 which was being shaded out. This tree, which had been continuously pollarded over the years, had a hollow trunk about 40 centimetres in diameter.[49]

Oral history handed down to Ian McLellan from older gardeners said that the original tree in Bed 84 was a cutting from 'Shakespeare's Mulberry'. Gardeners have long memories so I believed him, but a search of the early Annual Reports gave no clues. J. H. Maiden,

Director 1896–1924, went to Europe in 1900 but there is no record of him bringing back a cutting of Shakespeare's Mulberry. However, a bronze plaque was finally located in the Gardens Library with the following inscription:

Morus Nigra
(Mulberry)

> This tree originated from a Mulberry tree growing in Shakespeare's Garden, Stratford-on-Avon.
> Presented by the Shakespeare Society of NSW 18.4.34

The Annual Report of the Shakespeare Society for 1940 records:

> A Notable Event
> On Saturday afternoon May 25th, an interesting ceremony was held at the Botanical Gardens, when the Shakespeare Mulberry Tree was officially presented. This tree is a descendant of one which grew in Shakespeare's garden at Stratford-on-Avon . . .
>
> The President (Mr S.H. Bowden) . . . made the presentation to the Curator (Mr Hawkey) . . . Mr W. Lennard, M.A., gave an address on 'Patriotism in Shakespeare' (in the middle of World War II) and the Secretary spoke on the objects of the Society.

The Black or Common Mulberry, *Morus nigra*, a native of Persia is mentioned in the Bible. It has been cultivated as a fruit tree since ancient times, and lives to a great age, with individual trees that are known to have lasted at least 600 years.[50] The mulberry was cultivated in Roman gardens before the Christian era and may have been introduced into England in Roman times. It was an important tree in Medieval gardens. For a long time tradition had it that the first mulberry was planted at Syon (Sion) House, in London, in 1548 — although by this time it was a comparatively common tree in England.[51] Considered to be the 'wisest of trees', because it did not produce its new buds and leaves until the last frost was gone, the mulberry was used in heraldry as a symbol of wisdom.[52]

King James I is remembered for an initiative to try to establish a self-sufficient silk industry in England in 1606, and many thousands of plants were subsequently imported. Unfortunately these were the Black Mulberry, *Morus nigra*, which were not as satisfactory for feeding the silk-worms — which live on the leaves of the White Mulberry, *Morus alba*, from China.[53] The king planted four acres of mulberry trees on the site of what is now Buckingham Palace and gardens. This plantation became the celebrated Mulberry Garden referred to by Samuel Pepys and John Evelyn. Samuel Pepys did not visit the gardens until 20 May 1668, by which time they had deteriorated.[54]

There would have been plenty of mulberry trees in England in Shakespeare's time. In Act III, Scene ii of *Coriolanus*, Volumnia addresses Mememius Agrippa, and refers to the mulberry:

> Now humble, as the ripest mulberry
> That will not hold the handling.

Haydn's *Dictionary of Dates* indicates that Shakespeare planted a mulberry tree with his own hands in his ground at Stratford, and that Garrick, Macklin and others were entertained under it in 1742. Boswell records: 'the house, where Shakespear died, stood near the beautiful old Chapel in the middle of the town; and in the garden belonging, grew the famous mulberry tree'.[55]

Shakespeare's mulberry tree was later cut down by a clergyman named Gastrel, who had bought the house and was apparently sick of people coming to see the tree that had been planted by the bard. Charles Knight, in a note to his *Biography of Shakespeare*, says:

> [Shakespeare's] mulberry tree was cut down in 1756, and sold for fire wood, but the bulk of it was purchased by Mr Thomas Sharp, of Stratford, watchmaker, who, out of a sincere veneration for the memory of its celebrated planter, had the wood worked into curious toys and useful articles.[56]

There is a chair in the Shakespeare Room at the State Library of New South Wales reputed to be made from the wood of this tree.

What happened after the event of 1756 is pure speculation, but it is conceivable that people took cuttings and given the significance of the tree, this specific information could have been handed on orally through the generations. How a cutting of such a proposed cutting came to Australia is also unknown, but the Shakespeare Society of New South Wales connection gives credibility to the story. Mulberries had certainly been introduced into Australia as early as 1800.[57] Was one of them the celebrated cutting? We will probably never know.

On 23 April 1985 Shakespeare's 421st birthday/deathday (he died on his birthday) was celebrated in the Gardens, as part of a special joint activity held in association with the Friends of the Gardens and the Library Society. A special walk linking the Gardens with the State Library of New South Wales, researched and written by Kathleen Wall, Volunteer Guide, looked at the plants that Shakespeare wrote about. At the Gardens beginning of the walk the actor, Christopher Dibb, played Shakespeare and mused on the influence of his boyhood in the country; Coralie Tosswill played the lute in the Rose Garden and under the 'luscious woodbine'. At the State Library Neil Bonness, Rare Books Librarian, spoke of the man and his works, and a copy of the First Folio edition of the complete work of William Shakespeare (1623) was on display. It was an inspirational day and any subsequent celebration should include the Mulberry Tree in Lawn 17.

Shakespeare's Mulberry

Wisest of trees the mulberry
withholds its leaves beyond the hoar,
to vindicate some ancient law,
from Persia to the Roman See
as hostages to memory —
a mulberry planted by the bard
put Garrick, Macklin in its shade,
when both invited round for tea.

A mulberry planted as a poem
cut down for firewood or rind
a green twig kindling the clone
communion with its purple wine —
a quilted ransom for a king,
that will not stand the handling.[58]

1987
Garden: New Rose Garden
Planted by: See separate list below
Location: Lawn 15
Additional Information: The New Rose Garden, Lawn 15, Palace Garden grounds, around the existing pergola and pavilion (now called Rose Garden Pavilion) was designed by the Director, Professor Carrick Chambers, and sponsored by the Friends of the Gardens. The first plantings took place on 19 July 1987, the day of the presentation of the final instalment of the Friends' donation of $50,000 to the project. On that day, in beautiful sunshine, members of the Friends' Committee and representatives of the Executive, Trust, and Volunteer Guides planted climbing roses along the southern pergola as listed:

Western Side of Southern Pergola, South to North

Rose	Planted by
Rosa banksiae white	Allan Seale, Friends Committee
Rosa banksiae lutea (double yellow)	Toni Dingle, Friends Committee
New Dawn	Ruth Slade, Patrick Houlcroft, Gardens Staff
Rosa laevigata, Cherokee Rose	Joyce Beaumont, Friends Committee
Clair Martin (× 2)	Penny Farrant, Gardens Staff/ Allan Seale, Friends Committee
Lemarque (× 2)	Alan Tillsley, Gardens Staff
Madame Plantier (× 2)	Don Blaxell, Executive
Iceberg Climbing (× 2)	John Ferris, Trust Chairman/ Irene Morphett, Trust

Eastern Side of Southern Pergola, South to North

Rose	Planted by
Rosa banksiae alba plena	Bill Harrington, Friends Committee
Rosa bracteata	Toni Dingle, Friends Committee
Sparrieshoop	Flora Deverall, President Friends Committee
Cecile Brunner (× 2)	Joyce Beaumont, Friends Committee
Gloire de Dijon	Ron Tingle, Friends Committee
Golden Showers (× 2)	Boris Janes, Friends Committee; Margaret Spain, Friends Committee
Souvenir de Malmaison (× 2)	Lorna Lamy, Volunteer Guide Kathleen Wall, Volunteer Guide
Iceberg Climbing (× 2)	Allan Seale, Friends Committee

Ruth Slade and Pat Houlcroft dug the holes and prepared and tended the plants; the operation was supervised by Brian Donges. Professor Chambers was overseas at the time.

The garden was officially opened by Lady Rowland, wife of the Governor, on 10 April 1988. It was designed to show the rose in history, and the history of the rose

Choragic Monument in springtime, *Jaime Plaza.*

Thematic display at the Tropical Centre—a Bicentenary gift from BP Australia Ltd, *Jaime Plaza.*

Sunken Garden, *Jaime Plaza.*

in culture and cultivation, and the fascinating diversity of the genus *Rosa*, as expressed in the following verse:

The Rose

Crusading beauty, attar of the East,
Damask of monastery and Moor,
from blowzy harem, Roman wedding feast,
the House of Lancaster, of York,
to Gallic passion, Josephine's retreat
into a Bourbon haze, the war
against the infidel, for Christ —
sweet innocence of tooth and claw.
The romance we espouse is fixed
in Heraldry, a Rosary of hips
from dog-rose or from rambling brier
when flesh-pink shall to gold aspire —
predicating love's release,
a rose for war, a rose for peace.[59]

Included in the plantings are species and old-fashioned 'heritage' roses. The modern rose is displayed in colour sequences — whites, pinks and reds on one side; whites, pale yellows and golden colours to the other. Climbing and pillar roses form part of the sequence. Another pergola, built on the northern side of the Pavilion in 1989/90, was officially opened on 8 April 1990.

As the box hedge (*Buxus japonica*) grows and the garden matures it will provide the visitor with both an educational resource and a haven of peace and beauty. The area of the Old Rose Garden (planted 1936) will be used to display herbaceous and aromatic plants and retain the Nellie Stewart Memorial.

Before 1936 there was a rose garden west of the Herbarium (1907) and in the Domain near the statue of Robert Burns (1904). In 1901 roses were grown at the site of the Sunken Garden (Memorial Garden to the Pioneers, Palace Garden Grounds).

1987
Tree: *Araucaria cunninghamii* (Hoop Pine)
Planted by: King Baudouin I and Queen Fabiola of Belgium
Location: Government House Grounds
Additional Information: Planted 30 November 1987, west of main drive.

1988
Tree: *Araucaria cunninghamii* (Hoop Pine)
Planted by: HRH The Prince of Wales and HRH The Princess of Wales
Location: Government House Grounds
Additional Information: One tree planted jointly on 27 January 1988, west of main drive.

1988
Tree: *Quercus robur* (English Oak) — Phillip Oak
Planted by: His Excellency Sir James Rowland, Governor of New South Wales
Location: Bed 68
Additional Information: An oak tree was planted in conjunction with the dedication of a wall on 28 January 1988, constructed from bricks from the foundation of Captain Arthur Phillip's house at Vernals Farm, Hampshire. The Phillip Wall (see p.79) was a gift to the city by Sydney businessman, Mr Geoffrey Cottee. The oak tree came from an acorn taken from an ancient tree on Vernals farm.

1988
Tree: *Araucaria columnaris* (New Caledonian Pine)
Planted by: The Hon. Nick Greiner, Premier of New South Wales
Location: Lawn 15
Additional Information: Planted near the Rose Garden on 9 September 1988.

1988
Tree: *Quercus robur* (English Oak)
Planted by: His Excellency Sir James Rowland, Governor of New South Wales
Location: Government House Grounds
Additional Information: Planted 15 September 1988, south of Government House. Grown from acorn fallen from trees planted by Capability Brown when Petworth Park, Sussex, was landscaped in 1752.

1988
Tree: *Araucaria heterophylla* (Norfolk Island Pine)
Planted by: Queen Beatrix of the Netherlands
Location: Government House Grounds
Additional Information: Planted 5 November 1988, near tennis courts.

1989

Tree: *Araucaria cunninghamii* (Hoop Pine)
Planted by: His Excellency Sir James Rowland, Governor of New South Wales, and Lady Rowland
Location: Lawn 34
Additional Information: Planted on 10 January 1989 to mark his retirement as Governor.

1989

Tree: *Araucaria cunninghamii* (Hoop Pine)
Planted by: Mrs F. Kerr (President) and Miss M. Drake (Secretary), Sydney Garden Club
Location: Lawn 2
Additional Information: Planted on 7 March 1989 by the Sydney Garden Club to commemorate the long association of the Sydney Garden Club with the Gardens.

1989

Tree: *Araucaria luxurians*
Planted by: Mrs Joan O'Brien, President of The Garden Club of Australia
Location: Lawn 27
Additional Information: Planted on 27 September 1989 in conjunction with the Club's biennial Conference. The planting of a large number of *Araucaria* over the last couple of years was a conscious decision by the Director to re-establish the nineteenth century 'character' of the Gardens, as created by a large number of such trees.

1990

Tree: *Pritchardia maideniana*
Planted by: The Hon. Nick Greiner, Premier of New South Wales, and Mrs Greiner
Location: Lawn 24 (Band Lawn) — subsequently moved to Bed 55, near Gardens Restaurant.
Additional Information: Planted on 26 February 1990, on the occasion of the hundredth meeting of the Royal Botanic Gardens and Domain Trust.

The species was described in 1913 from two mature specimens growing in these Gardens. It has not been found in the wild, despite extensive searching, and the source is unknown. *Pritchardia* is a genus of some 37 species, all but four of which are Hawaiian. The rest are from Fiji, Tonga and the Danger Islands. Many species are rare and endangered, due to the predation of their fruit by rats and feral pigs.

1990

Tree: *Adansonia gregorii* (Baobab)
Planted by: Ms Wendy Machin (MP for Manning) and Mr Michael Photios (MP for Ryde)
Location: Outside the eastern end of the Arc Glasshouse. Another tree was later planted inside the Pyramid Glasshouse.
Additional Information: The baobab tree outside the Arc Glasshouse was planted on 17 October 1990. The three metre-high trees from a Broome nursery were flown in a Caribou aircraft from Derby Air Base to Darwin where they were loaded on to a Hercules plane and flown 3000 km to Richmond Air Base. The airlift was arranged by Gardens Media Liaison Officer, Joe Glascott, with the co-operation of Wing Commander Jack Edmunds, Commander of RAAF for Northern Australia.

1990

Tree: *Araucaria montana*
Planted by: His Excellency Rear Admiral Sir David Martin and Lady Martin
Location: Lawn 30
Additional Information: Planted 4 August 1990 to honour the ninetieth birthday of Her Majesty Queen Elizabeth, the Queen Mother, Patron of the Victoria League for Commonwealth Friendship. Sir David died a few days later.

1991

Tree(s): *Eucalyptus robusta*
Planted by: (In order west of Lion Gate): His Excellency Rear Admiral Peter Sinclair, AO, and Mrs Sinclair: Mr M. A. Besley, AM, Chairman Royal Botanic Gardens and Domain Trust; Mrs Sally White (representing the [Vincent] Fairfax Foundation); Mr Peter E. J. Jollie, P & O Australia Ltd; Mr Peter Euston and Mrs Lorraine Amos, descents of John Richardson, the 'gardener's boy' who assisted Mrs Macquarie replant the original Wishing Tree; and Professor Carrick Chambers, Director Royal Botanic Gardens.
Location: Lawn 26, north of Macquarie Wall
Additional Information: Planted Sunday 16 June 1991, as replacement trees for avenue planted in 1816, as part of our 175th anniversary celebrations. The actual anniversary was celebrated on Thursday, 13 June 1991.

4

The Public Works

Alphabetical Checklist of Existing Buildings, Walls, Pavilions, etc. in the Gardens and Domain

Anderson Building (Old Herbarium — also see Brown Building)

Date built: 1878/1899
Additional Information: Two-storey additions to earlier Museum and Lecture Hall built in 1878, in Italian modified style to harmonise with the old museum. The first stage of the building, designed by James Barnet, Colonial Architect, was completed in 1878. It was a single large room, of rendered masonry construction, and with a steeply pitched slate roof with a central highlight. The second stage of the building, built by J. M. Pringle of Waverley for £5,766, was completed in 1899. It included three large rooms for the Herbarium, a Library and Office, Director's Office, Assistant's Office and Superintendent's Office, Museum and two seed rooms.

The Old Herbarium was substantially modified during the construction of the Robert Brown Building, and now contains the Visitor Centre and J. H. Maiden lecture hall. The upper level contains the Director's Office and various other offices.[1]

Aquatic House (see Cryptogam House)

Art Gallery of New South Wales

Date built: Commenced 1896
Additional Information: The New South Wales Academy of Art, founded in 1871, was the forerunner of the Art Gallery of New South Wales. The art works for the Sydney International Exhibition had originally been allocated space in the galleries of the Garden Palace. However, space became limited, and the light quality in the Garden Palace was considered unsatisfactory. The collection then occupied a building, designed by William Wardell, located just to the west of the main Botanic Gardens entrance from 1879 to 1885. Fire risk, damp and termite damage to this building sharpened the move to establish the Art Gallery.

A new building, designed by Horbury Hunt, was erected on the present site and cost £11,000. Between 1896 and 1909 the Government Architect, Walter Vernon, added the southern wing, the façade and portico, and a northern picture gallery. The north-west wing cost £9,564 (tendered by Mr C. Harris) and was opened to the public early in 1909.[2]

The north-eastern extension was completed between 1968 and 1972. The eastern extension, designed by Andrew Andersons, Government Architect, was opened to the general public in 1988 and received the Sulman Award for Architecture in 1989.

The various bas-reliefs around the front of the building — projected as part of an unfinished series — were completed as follows:

Name: Assur-Natsir-Pal, King of Assyria
Artist: Gilbert Bayes
Date: 1807

Name: Queen Hatasu of Egypt
Artist: Countess Feodora Gleichen
Date: 1808

Name: Phyryne before Praxiteles
Artist: Percival Ball
Date: 1900

Name: Augustus at Nimes
Artist: William Reid Dick
Date: 1931

The Museum and Lecture Hall (site of current Visitor Centre) constructed in 1878, *RBG Library*.

(Robert) Brown Building (Herbarium — also see Anderson Building)

Date built: Completed 1982

Additional Information: Until the appointment of J. H. Maiden as Director in 1896, there was 'practically no herbarium in existence at the Botanic Gardens'. In his previous position as Curator of the Technological Museum Maiden had accumulated a large number of herbarium specimens for scientific study, and one of his first priorities on taking office at the Gardens was to build up a National Herbarium in Sydney.[3]

After the death of Sir Joseph Banks in 1820 little interest was taken in Australian botany until the appointment of Sir William Hooker as the Director of the Royal Botanic Gardens at Kew in 1841, who tried unsuccessfully at the time to establish closer relations between Sydney and Kew.[4]

The distribution of duplicates of Robert Brown's collection took place in 1876. (Brown was the famous English botanist who sailed with Matthew Flinders.) Sydney was offered one of these sets but the offer was not taken up.[5] Earlier, Sir Thomas Mitchell had given a collection of sub-tropical Australian plants to The Australian Museum in 1852. It should be noted in this context that The Australian Museum and the Botanic Gardens were under the same body of Trustees from 1836 to 1851.[6]

In his evidence before the Select Committee on the Botanic Gardens in 1855, Charles Moore stated that he had received nothing from his predecessors, and that the Botanic Gardens were 'utterly destitute' of Herbarium or Library when he became Director.[7] It is therefore apparent that the Mitchell collection did not come under the care of Moore, but that at a later date Ferdinand von Mueller returned some of these specimens to the Gardens via Sir William Denison.[8] Mueller had visited Sydney in 1856 and these plants may have been returned as a result of this visit.

In 1846 Ludwig Leichhardt (who had unsuccessfully applied for the position of Superintendent of the Sydney Botanic Garden in 1844) exhorted Parisian friends 'to work harder (for) a herbarium, for Sydney'.[9] Daniel Bunce, plant collector on Leichhardt's 1846 expedition,

J.H. Maiden at his desk in his office, *c.* 1900, *RBG Library*.

Museum established by Charles Moore, *RBG Library*.

had also unsuccessfully applied for the position of Superintendent of the Sydney Botanic Gardens. Bunce was to have 'an equal share of seeds, and plants where triplicates were available and could be carried. All unique specimens, and all of which only duplicates could be taken, belonged to Leichhardt, or rather to the future Sydney Herbarium'.[10]

On 10 June 1856, Moore, referring to his Botanical Library and Herbarium, noted a

> considerable utility of those interested in [the] science of botany, continue steadily upon the increase. The latter, in particular, has been considerably enlarged, by donations from the Governor General and Dr Muller [*sic*], of types of the flora of Victoria, and by a choice collection of Algae from the coasts of New Holland, presented by Professor Harvey. To these the public have free access three days a week, between the hours of 10 am and 5 pm.[11]

In 1854 Charles Moore reported that 'a Collection of Specimens of plants, principally indigenous to this Colony, is in the process of being arranged in books . . . The Ferns have already been completed as far as possible.'[12] The Public Botanical Library (embryonic Herbarium?) had been established in 1853.[13]

In the summer of 1857, in 'view of rendering the Public Herbarium of indigenous plants as complete as possible', Charles Moore went to the Blue Mountains and travelled along Bell's Line of Road as far as 'Mount Thomar' (*sic*) where he remained for some days. During this journey he obtained 'flowering specimens and seeds of many plants new to the collection, and also of some previously not known. By this means the Herbarium has been greatly enriched'.[14]

In his estimates for 1858 Charles Moore allocated £50 (a lot of money in those days) 'towards the formation of a Public Botanical Library'. In a collection of letters held in the State Archives, one from a Mr A. Thuzet of Rockhampton, dated 3 March 1859, states, 'The Sydney Museum (site of current Visitor Centre) will soon put before the public an Herbarium'.[15] The word was obviously out that a Botanic Library and Herbarium had been established at the Sydney Gardens.

By 1870 Charles Moore felt confident enough to say that 'Valuable additions [had been] made to the Library,

The National Herbarium of N.S.W., *Government Printing Office collection, State Library of New South Wales.*

which now contains many of the best standard works on Botany, and also to the Herbarium. Both of these afford a ready and most useful means of reference to those who desire to consult them'.[16]

When J. H. Maiden became Director in 1896, he told Mr Harwood, Superintendent of the Botanic Gardens for many years, that there had been a set of Allan Cunningham's plants in the old cottage which Moore left early in 1875. These specimens were afterwards found by Maiden in the old seed room, to which they had been removed when the cottage had been demolished. On closer examination these turned out to be the remains of the lost Mitchell collection. Of all the collections of all the earlier explorers, this is all that remained.[17] Apparently the earlier explorers had regarded their collections as their own private property and these were often sold on their death by their heirs.

There are currently some Mitchell specimens in the Herbarium collection, on smaller sheets of paper, with both field labels and printed labels. These are most probably the original specimens.[18] The Sydney Herbarium also contains Leichhardt specimens which came from his executor. Leichhardt's will had bequeathed his specimens to The Australian Museum and the Sydney Botanic Garden.[19] Although he wanted his material to come to Sydney, a large part of the collection went to Ferdinand Mueller in Melbourne in 1861.[20]

Most of the material collected when Charles Moore as Director was sent to Kew. As the collector's name is often not on these specimens, the handwriting needs to be compared to see if the collections were made by Moore or by William Carron who survived the Kennedy expedition to Cape York (1848) and was later to work as a collector for the Sydney Gardens.

As well as the Carron/Moore material the Herbarium has specimens collected by Robert Brown, Allan Cunningham, George Caley, Charles Fraser and others. Most of these came from duplicates held in the British Museum (Natural History) in more recent times.[21]

During Maiden's visit to Europe in 1900, he visited botanical museums and established personal contact with curators. Then, through the cooperation of Mr Britten, the Trustees of the British Museum presented Maiden with nearly 600 duplicates of specimens collected by Banks and Solander at Botany Bay in 1770.[22]

When Maiden became Director, there was the single storey Museum and Lecture Hall (now the site of the Visitor Centre and Gardens Shop) that had been built in 1878. 'Beyond the small greenhouses, we catch a glimpse of the stonework being prepared for the new buildings, which are to contain Museum, Library, Lecture Rooms, and other arrangements for promoting botanical research.'[23]

First-floor additions were made to this building in 1898, providing three large rooms upstairs for herbaria, Director's office and an office for an assistant. During the rebuilding process most of the original (1877) building was taken down, retaining only the walls of the Museum and a portion of the back with the seed room.[24]

The architectural style of the new building, with its slate roof, designed by the Government Architect, W. L. Vernon, was 'Italian modified' to harmonise with the old museum. A portico was introduced, forming an entrance to the National Herbarium, and to balance the design. The interiors today are little altered, retaining their cedar skirtings, staircase and doors; zinc ceilings and cornices, and many other fittings.[25]

The Herbarium was opened by the Chief Secretary, the Hon. John See (later Sir John) on 8 March 1901.

> The building is a beautiful one, and worthy of its charming surroundings. The exterior has called forth the admiration of visitors, while its internal arrangements are most convenient, the architect having consulted my wishes in every possible way. Although the building is of considerable size, because of the architecture adopted, its situation and arboreal surroundings, it is not unduly conspicuous, thus not dominating or overpowering the Gardens in any way.[26]

The mood of the period is captured in the following poem by Kenneth Slessor (although I note the definite article is missing in the Roman capitals on the portico). Nor do we have black swans in our pond these days — but the poem is definitely Sydney.

Elegy in a Botanic Gardens

The smell of birds' nests faintly burning
Is autumn. In the autumn I came
Where spring had used me better,
To the clear red pebbles and the men of stone
And foundered beetles, to the broken Meleager
And thousands of white circles drifting past,
Cold suns in water; even to the dead grove
Where we had kissed, to the Tristania tree
Where we had kissed so awkwardly,
Noted by swans with damp, accusing eyes,
All gone today; only the leaves remain,
Gaunt paddles ribbed with herringbones,
Of watermelon-pink. Never before
Had I assented to the hateful name
Meryta Macrophylla, on a tin tag.
That was no time for botany. But now the
 schools,
The horticulturists, come forth
Triumphantly with Latin. So be it now,
Meryta Macrophylla and the old house,
Ringed with black stone, no Georgian Headlong
 Hall
With glass-eye windows winking candles forth
Stuffed with French horns, globes, air-pumps,
 telescopes
And Cupid in a wig, playing the flute,
But truly, and without escape,
THE NATIONAL HERBARIUM,
Repeated dryly in Roman capitals,
THE NATIONAL HERBARIUM.

By the 1950s Maiden's Herbarium was becoming crowded with its collection, approaching one million specimens, stacked in boxes from the floor to the 14-foot ceilings. R. H. Anderson, Director 1945–64, fully aware of the problems and needs, strove to obtain a new building but was frustrated by the times.

The new Herbarium and Library, called the Robert Brown Building, was not completed until 1982, during the Administration of Dr L. A. S. Johnson, Director 1972–85. The funding was achieved through the good services of both Sir Alexander Beattie, Trust chairman at the time, and the Hon. Neville Wran, Premier of New South Wales.

The Robert Brown Building, a contemporary unobtrusive flat-roofed style of construction using poured concrete and glass, links both the old Herbarium (Anderson Building) and the former Director's residence (Cunningham Building) and is a veritable palace for plants. The project architect was Brian Zalaikha and the supervising architect was M. Fletcher from the Special Projects Branch of Government Architects. Documentation was by Ancher, Mortlock and Woolley and the builder was James Wallace Pty. Ltd. The cost was $5.1 million.

The move to the Robert Brown Building gave the opportunity for a major reorganisation of the specimen collection using the Dahlgren System (a Danish botanist's way of arranging the major groups of flowering plants) and specimens were transferred to new plastic boxes.

Chess Pavilion (see Rose Garden Pavilion)

Cunningham Building (Director's Residence)

Date built: 1874
Additional Information: Formerly the Director's residence, a two-storeyed asymmetrical Italianate house, probably designed by James Barnet, Colonial Architect. It is constructed of rendered brick, stuccoed, with a tower and ground-floor verandahs. The style was 'invented' by Prince Albert for the building Osborn in the Isle of Wight, and became very popular in the then Empire. Australia has a number of such buildings, both public and domestic, and this building is a fine example of the type.

R. H. Anderson was the first Director/Curator to choose not to live on site, and remained in his house at Chatswood.

In 1969 the Cunningham Building was converted into administration offices and restored. In 1983, after the completion of the Robert Brown Building, it was used by the Living Collections Division and Horticultural Botanists who moved to the new building in the Depot in 1990. After that the downstairs part of the building was occupied by Community Relations (Public Relations, Editorial, and Interpretation) and this book was drafted in what was J. H. Maiden's old Library.

Conservatorium of Music

Date built: *c.* 1821
Additional Information: The Conservatorium of Music, designed by Francis Greenway, was originally built as a castellated stables and servants quarters for a proposed new Government House. Commissioner Bigge, in 1819, rebuked Governor Macquarie for building 'such a palace for horses', but did not stop the construction. The stables were used as such by successive Governors until the coming of the motor car at the beginning of this century. They were roofed in 1908 and W. L. Vernon's alterations, commencing in 1908, were finished in 1911.

In 1912 a Committee was set up to establish a Conservatorium of Music within the walls of the old stables. The inaugural concert was held on 6 May 1915.[27]

The music school (once part of Sydney Teachers' College, then Alexander Mackie CAE, and now the University of Sydney) offers tuition in 'big band' jazz, and classical. Consequently their music and songs add a special ambience to the western side of the Gardens. Young people sometimes play their instruments in the open, and are commissioned to play for the Gardens' Spring Festival and other special occasions.

Cottage 2 (see Lion Gate Lodge)

Cottage 4 (Herbarium Seed House)

Date built: 1863
Additional Information: This two-storeyed stone cottage, on the eastern side of the Depot, with timber verandahs with slate roof, was built as a residence for the Superintendent or Head Gardener, extended in 1900, and renovated in 1922, 1972 and 1984. It was then used as part of the Herbarium and temporary seed house until 1984 modifications, when again it was used as a residence.

From an old photograph supplied by Daphne Robilliard, I believe her great grandfather James McFarland, Head Gardener, lived in this cottage *c.* 1875.[28] After an inspection with Alan Tillsley (who lived in Cottage 4) in December 1990 — and because of the position of the windows (and especially a gable window dotted into the *c.* 1900 plan), and the pitch of the roof (with its original slate) — I am convinced that this photograph shows the eastern side of what is now Cottage 4.

The Gardens' Curator Edward Ward lived in Cottage 4 from about 1914 to 1935. In those days the house was entered from the Domain and had a vegetable garden and a tennis court on the western side.[29]

Cottages 5 and 6 (see Twin Cottages)

Cryptogam House (Aquatic House)

Date built: 1921

The Depot, 1915, *RBG Library*.

Additional Information: The old Aquatic House (Cryptogam House) — a neat stone building in Bed 23 containing glass tanks displaying water plants — 'was opened to the public, without ceremony, on March 13th'.[30] The building was later used to grow and display cryptogams — mosses and ferns, etc. — and was used by Betty Jacobs, Education Officer, in the 1970s, but then fell into neglect. The building was renovated in 1987.

Depot and Nursery

Date built: —
Additional Information[31]: The site of the Depot and Nursery was formerly the Governor's Kitchen Garden. The earliest structure still standing in this area is the first stage of Cottage 4, constructed in 1863. There have been many changes over the years, with new glasshouses being erected on the base courses of earlier ones.

The first building in this area, other than the residence and glasshouses, was an L-shaped stables building which is first indicated on the 1879 plan of the Gardens. This has now been partially demolished, the remaining section having been extended to become the carpenter's workshop. The Annual Report of 1898 states that:

> Packing and potting sheds and store-rooms, together with a kitchen and dining room for the men, have been erected during the year. The packing shed is a continuation of the existing stable building in an easterly direction, and a room, 35 feet × 18 feet, is provided, together with a loft for packing material. Great saving of labour and economy of time have resulted through the situation and the arrangement of this building. Parallel with this, and at a distance of 48 feet from it, is the building which contains a lofty potting shed, 23 feet by 13 feet 6 in.; two storerooms, 11 feet ×13 feet 6 in. respectively; a men's kitchen, 11 feet ×13 feet 6 in., adjoining which is a room 19 feet ×13 feet 6 in., which is used as a dining and recreation room. It is plainly but sufficiently furnished, and is much appreciated by the staff. The parallel lines of buildings are joined at their north-easterly and south-easterly corners by a substantial fence, thus forming a quadrangle.

The following other insights are provided by David Churches:

> Tool House and Motor Mower House — a brick building 36 feet × 20 feet has been erected. Two-thirds of this building is now used as a tool house for outdoor staff and

gardeners and labourers, each man having a separate locker for his tools. Hitherto tools had to be stored in isolated wooden buildings which had done service as ticket offices during the Sydney International Exhibition of 1879–80.

The Motor Mower House takes up the remaining portion because additions to the Refreshment Rooms rendered it necessary to pull down the old wooden shed formerly used as a machine shed.

New Depot Offices — a new brick building has been erected near to the Tool and Motor Mower Houses, and comprises Timber Shed, Painter's Shop, Label-writer's Room, Forage, and Police Room.

The old unsightly and dangerous timber shed with its open front was demolished.

Men's Dining Room — the dining room and kitchen have been re-arranged by knocking out a partition wall, and so making provision for seating forty persons instead of twelve.

Managements for the supply of hot water for tea, etc., have been transferred to a boiler-house which has been built at the end of the Tool House.

Recent changes in the Depot include the construction of a somewhat unsympathetic roof over the area between the carpenters' workshop and staff amenities to form a larger workshop, demolition of the incinerator, and construction of a women's amenities block adjacent to the men's change room and showers.

Director's Residence (see Cunningham Building)

Domain Lodge (St Marys Lodge)

Date built: 1835
Additional Information: The Domain Lodge, at St Marys Gates, was designed by Mortimer Lewis and built in 1835. Tenders were accepted in May 1835 from Messrs Ross Coulter & Co for £290 10*s.* 1*d.* An application in June by the contractors to quarry stone in the Domain was refused.[32]

Drawings by Conrad Martens and John Skinner Prout show the Lodge in its original form (prior to 1845), without the eastern or northern extensions and with quarry glazed windows and shingle roof.[33]

In 1853 there was a wooden shed at the rear of the Lodge which had 'been built for many years and [was] in a state of dilapidation'.[34] By 1856 water had still not been connected to the Lodge.[35]

The Lodge, western entrance to the Domain, *David Churches.*

The windows of this Lodge were smashed on 16 December 1856, as the unfortunate Frenchman M. Maigre had been hiding there after a visit in the Domain when his balloon had failed to ascend.

The ceilings were re-canvassed and papered in 1866, and in 1907 an outside stable was converted into a bedroom opening from a kitchen, and a large wooden shed built in the yard for stores, tools etc., at a total cost of £76.[36]

Once constructed, the Lodge came under the control of several departments. The first occupant was a police constable whose name has not been found. When Mortimer Lewis was appointed Honorary Ranger of the Domain in December 1836 he recommended James Lackey as Bailiff. Lackey lived in the Lodge from 1837 to 1849. The occupants from 1849 to 1851 are not known. The following people have lived in what is one of the oldest, if not the oldest continually used residence in this part of Sydney, as evidenced by the huge *Euphorbia* in the Gardens.[37]

1851	James Breuster	Bailiffs
1852	William Johnston	
	William Gibbons	
1855	George Kemp	
1864	John Meads	
1875-1904	Finlay McKay	
1904-41	(not known)	
1942	J. R. Whitacre	Superintendents
	J. T. Driver	

The Palm House with National Herbarium (Anderson Building) in background, 1915, *RBG Library*.

1951	H. Grant	Foreman
1953	J. Thomas	Assistant Superintendents
1968	Barry Dangerfield	
1977	C. Weale	
1980	Barry Dangerfield	Superintendent
1982	Geoff Senior	Ranger

Fleet Steps

Date built: 1908
Additional Information: Built as part of the celebrations associated with the arrival of the Great White American Fleet in 1908. The Governor-General landed at this site in 1901 as part of the ceremonies associated with the inauguration of the Commonwealth of Australia. Queen Elizabeth II landed at this same spot in 1954, as recorded in the commemorative wall.

Gardens Restaurant (see Kiosk)

Glasshouses

Date Built: 1876 to present
Additional Information: All the glasshouses in the Depot have been rebuilt in recent years because of the deterioration of the original steel-framed structures.[38]

Several of these glasshouses were originally open to the public as exhibits. This continued until the mid-1970s, when the completion of the Pyramid Glasshouse allowed them to be converted for propagation uses only.

Glasshouse No. 10 (The Palm House/Education Glasshouse), probably designed by James Barnet, Colonial Architect, was erected in 1876 and reroofed in 1899 and 1912.[39] It was called the Orchid and Show House in Maiden's map of 1903.

Glasshouses Nos. 1 and 2 were constructed prior to 1894. Glasshouse No. 1 was used for tropical ferns and Glasshouse No. 2 for orchids. These glasshouses 'received new doors and stages' in 1899, and had new heating installed in conjunction with the construction of Glasshouse No. 3.

> A low roofed orchid house, 40 × 15 feet, has been erected. No wood has been used in its construction except for the doors, the benches being of wrought-iron. The pitch of the roof has been kept as low as it was possible to construct it, and the panes of glass both of roof and sides are of very large size. This is a departure, at all events as regards Australia, and up to the present no sheets of glass have been broken, although the house has been in use for several months. In fact, the house appears to be quite satisfactory. It was erected by Mr J. M. Wilson at a cost of £426. It is intended to be used as an Australasian and Polynesian orchid house; but as our collection of such plants requiring the protection of glass is not sufficient at present to fill it, it is temporarily used as a miscellaneous house.[40]

Glasshouse No. 7, originally called The Orchid House, was constructed in 1899 to the design of Walter Vernon, Government Architect. It was built to house a collection of Australasian orchids donated by Mr Hugh Dixson.[41]

Glasshouse No. 3, which was originally called the Fern House, was constructed in 1900 under the direction of Walter Vernon, Government Architect. In 1899 it was recorded that: 'Another iron hot house uniform in design with the two which are presently devoted to orchids and to miscellaneous plants respectively has arrived from England'.[42] The following year: 'This house has now been erected and has been in use for some months. It is in fact, crowded already, as all our houses are, elequent testimony to the way in which our collections are inceasing.'[43]

Glasshouses Nos. 5, 6 and 9 were originally constructed in 1909, also under the direction of Walter Vernon.

> Three additional plant houses have been erected by the Government Architect's Department. They are substantially built, of neat design, and are well adapted for the purpose required. As soon as the houses were completed they were

> immediately stocked with plants to their utmost capacity; thus the overcrowded state of the other houses was greatly relieved. Also, in connection with the new houses above mentioned, a spacious boiler-house, with improved hot-water service, has been provided.[44]

Glasshouse No. 4, which was originally called the Show Pavilion, was erected in 1911, once more under the direction of Walter Vernon.

> A new heated conservatory or show house, 50 feet × 30 feet, has been erected and was opened to the public in August. This house was much needed for the display of flowers, palms, ferns and other interesting plants; the floor space being ample, the contents of the house can be viewed in comfort since there is no centre stage. The public have been warm in their appreciation of this house.[45]

Glasshouse No. 8, originally intended for biological research, was constructed in 1923 to the design of G. L. Blair (Government Architect).

> A new glass house for biological research work has been built together with a soil sterilizing and potting shed. These will be used by the biologist of the Department of Agriculture. The glasshouse has been built in a solid manner, with safety doors to ensure, as much as possible, proper control of any disease which may for a time be under observation. It has been placed near the main block of glasshouses, so that the same hot water heating arrangements can be supplied without extra cost in service or attendance. The cost of this work was within the contract price of £1,010.[46]

All of these glasshouses have now been substantially modified since they were originally constructed. More recent glasshouses include those constructed in 1961 and 1965. Since then two more glasshouses have been constructed near Cottage 4 — one on the area previously referred to as the shade house, the other being the quarantine glasshouse.

What is termed the Fernery on the 1978 plan was called the Bush House in Maiden's map of 1903. It was constructed at some date between 1870 and 1894 — '75 feet long by 42 feet wide by 30 feet high, with a semi-circular top and covered with Tea Tree (*Leptospermum flavescens*, sm). The Bush House contains many valuable tree-ferns and palms'.[47]

Alan Tillsley told me the Fernery (constructed from the remains of the Governor's Bathing House in 1924) was open to the general public for Spring and Summer shows until the late 1960s. The pathways were of red gravel, and there were two ponds and a stone bridge. The rockery was constructed of the same conchoidal stones as those found in the rockery under the Port Jackson Fig, near Glasshouse No. 10 (Whale's Head Rockery) and may also have come from Little Bay in the 1920s. Similar stones were also used in landscaping near the Boy with Thorns statue.

The semi-circular roof of the Fernery was later replaced by a trussed roof with corrugated fibreglass cladding.

Construction of a new Fernery, made possible by a generous donation following a visit to the Gardens, commenced in March 1992 on the site of the old Fernery. The award-winning design was by the then architect student, John Barbacetto.

(Tropical) Glasshouses (Sydney Tropical Centre — The Arc and the Pyramid)

Date built: Opened 1990
Additional Information: The Pyramid Glasshouse, designed by E. H. Farmer, Government Architect, and constructed by E. A. Watts Pty Ltd for a cost of $112,000, was completed in 1972.

One of the first of its kind in the world, the unique Pyramid Glasshouse was built of a triodetic aluminium frame with glazing bars and 6 mm toughened tinted glass. It was equipped with an automatic heating and cooling system, to maintain an internal temperature of 17–28°C and a relative humidity of 55–90%, and a circuit of walkways and elevated platforms from which to view the rainforest plants.[48] Limited funding delayed the completion of the project, and the glasshouse was not opened to the public until 1 July 1976.

In the early 1980s a public appeal was launched (The Great Pyramid Appeal) to obtain the money to build two more pyramid glasshouses, at the then estimated cost of $1.2 million. This appeal was later aided by a substantial donation from the industrialist and orchid lover, G. Hermon Slade.

The revised project, assisted with financial support from the N.S.W. Government, BP Australia Pty Ltd, the N. & H. Slade Trust, and public donations to the Glasshouse Appeal was costed in excess of $4 million. The revised design by Ancher, Mortlock & Woolley, featured a quarter-circle 'arc' (to allow for extended internal height of plantings) with the existing pyramid at its 'focus'. The Sydney Tropical Centre (comprising the Arc and the Pyramid linked with an underground interpretive space) was built by Reed Constructions Pty Ltd for the Public Works Department, and was the overall winner of the Excellence in Construction Awards, 1989, of the Master Builders Association of New South Wales and the *Daily Telegraph*.

A new road system was installed by Gardens' staff and the whole complex, including a crescent of outside rainforest plantings, was opened by the State Governor, His Excellency Rear Admiral Peter Sinclair, on 16 September 1990.

Government House

Date built: Occupied 1845
Additional Information: In 1810 Governor Macquarie commissioned Francis Greenway to design a new residence — a castellated House of Stone — but only the stables (which now form the nucleus of today's Conservatorium of Music) were built. In 1827, during Governor Darling's term, the site for the present house was chosen and plans called for, but the project was discarded.

It was not until 1835, after further protests from Governor Bourke, that the plans were prepared in London, site unseen, by the architect, Edward Blore (1789-1879), who was special architect to King William IV, and later to Queen Victoria.

Construction began in 1837 under the supervision of Mortimer Lewis, Colonial Architect. The house, a fine example of early Victorian Gothic revival, was not occupied until June 1845 by Governor Sir George Gipps.

Government House and garden now occupy five hectares within the Royal Botanic Gardens, near the Sydney Opera House and overlooking the harbour.

Government House Lodge

Date built: 1835
Additional Information: Government House Lodge was built in 1835, the same year as Domain (St Marys) Lodge. A stone windmill erected on the site and owned by a Mr Steele was ordered to be demolished and the stone used to erect this new lodge — implying there had been an old one. A tender was accepted in May 1835 from Messrs James Fergusson & Co. for £319 1*s*. 0*d*.[49]

This Lodge features in a watercolour by John Rae, dated 1842. It was designed in the 'Gothick' style by

The Gardens Restaurant, *c*. 1943, *Mrs Peg Deane*.

Mortimer Lewis, and similar to the one at the entrance to Royal Windsor.

Originally virtually due west of the stables (Conservatorium) as shown in a map of 1894, it was relocated in 1916 when Macquarie Street was widened. A new Gate House was built in the 1930s, and the carriage road and gates to Government House were realigned to the east of the Lodge in the mid-1960s, after the construction of the Cahill Expressway.

Herbarium (see Anderson Building and (Robert) Brown Building)

Kiosk (Refreshment Room/Gardens Restaurant)

Date built: 1886
Additional Information: The original Kiosk (Refreshment Room) was built in the Gardens in 1886, designed by James Barnet, Colonial Architect, at a cost of £1,252 1*s*. 0*d*.[50] The building was repaired in 1897 at a cost of £124 3*s*. 4*d*.;[51] the drainage was improved in 1901,[52] and additions were made in 1906, 1908, 1911, 1915 and 1952.[53] In 1952 a single storey extension was added to the north of the building, providing a take-

away facility, with an open terrace on its roof providing additional seating for the Refreshment Room.

In *Water Under the Bridge*, Sumner Locke Elliott, describes a childhood visit to the Refreshment Room at the Sydney Botanic Gardens in the 1930s. The character disliked outings in the Gardens because:

> There was a sense of gloom where the giant fig trees [*sic*] met overhead and overshadowed the glass-roofed kiosk where he had been reluctantly led to have an ice cream at the fly-spotted marble tables. The green light made it spooky, predatory pigeons eyed him through the dirty windows.
>
> In between the stifling bamboo groves there were stagnant reeking ponds where under the water lilies sinister fins moved. The statuary was abysmal, with stone nymphs missing noses and fingers, and pugnacious boxers in fig leaves faced each other across paths, grottos that should have enticed led implacably to rest rooms.
>
> On the hard benches the unemployed lolled reading newspapers they retrieved from rubbish baskets, and from the nearby Conservatorium of Music the continuous chanting of scales came to seem after a time like the wailing of the damned.[54]

On the night of 19 March 1976, the Kiosk was destroyed by fire, with only the masonry façade of the original structure left standing. Fortunately this wall (dating from the 1880s) was its most attractive architectural feature, and has been retained in the new building.

The Gardens Restaurant, constructed in 1978 for a cost of $356,000, combined the qualities of a traditional garden pavilion and modern cafeteria-restaurant, with a Kiosk underneath. The project architects were Andrew Andersons and David Churches, design and development was by Ancher, Mortlock and Woolley, and the builder was D. G. Sundin & Co Pty Ltd.

The 1914 map of the Gardens also indicates that there was a Kiosk in the Domain across what is now Art Gallery Road from the site of the old bandstand (current site of Restaurant on the Park).

(Garden Palace) Kiosk

Date built: 1988
Additional Information: This small kiosk in Macquarie Street near the Palace Garden Gates was built as a bicentennial project as part of the Macquarie Street redevelopments.

Lion Gate Lodge (Cottage 2, Middle Lodge)

Date built: 1878
Additional Information: 'A new house has recently [1878] been constructed on the eastern side of the garden for the Overseer.'[55] 'A neat little house . . . has just been completed for Mr Duff, the Acting Head Gardener' at the end of the animal department (Zoo).[56] Called Cottage 2, it was referred to previously as Middle Lodge. Additions were made to the Cottage in 1907 and prior to 1921.

James (Jim/old Jim) Stovold, lived in the cottage from *c.* 1932 to 1946. One of his jobs was to lock the gates after the (central) bell rang at sunset.[57] He did this accompanied by his Airedale dog. The last gardener to live there was Ken Rendall, Foreman.

In 1987 the name was changed to Lion Gate Lodge (after the lion and lioness statues near the gate) when it became the residence for the Trust and Friends of the Gardens. The gardens around the Lodge were landscaped in 1989.

Macquarie Wall and Gate

Date built: 1812
Additional Information: The stone-faced and capped rubble wall dividing the old garden (Middle Garden) from what was once part of the Domain (now Lower Garden), erected *c.* 1812, is all that remains of the sandstone walls of the Macquarie era.

The Cartwright plan does not indicate whether the portion of the northern boundary constructed on the western side of the creek was of palings or stone. The 1833 plans, however, suggest that the sandstone wall was constructed only on the eastern side of the creek.

The gate in the middle of this wall is not shown in the Cartwright plan, but it appears in plans dated 1822 and 1833. A gate is shown at the extreme eastern end of the wall in the 1867 plan of the Gardens. The current eastern gate, which is marked on the northern side by the Lion and Lioness, was cut through the original wall in 1901. The stone piers came from the original Victoria Lodge Gate (erected in 1864).[58]

By the end of 1970 the wall was in poor condition, falling over, and covered with the rampant exotic fig, *Ficus pumila*. The stone-capped piers to the gateway remained, but no gates are known to have survived. In 1975 and 1977, with some financial assistance from the

Sydney Garden Club, the climbing fig was removed and the wall was restored and partially rebuilt.[59]

The only remaining portion of the wall in an 'original' condition is the easternmost portion (near Lion Gate Lodge).

The foundation to this wall was partly renovated in 1989.

Maiden Memorial Pavilion

Date built: 1929
Additional Information: At a public meeting convened by the Royal Society of New South Wales on 12 August 1926, it was decided to perpetuate the memory of Joseph Henry Maiden by erecting a suitable memorial in the Gardens. Subscriptions were sought and in 1929 the Maiden Pavilion, designed by R. M. S. Wells, Government Architect, was built on Lawn 25. It was dedicated by the Premier, Mr. Bavin, on 20 May 1930.

Mitchell Library (see State Library of New South Wales)

Opera House

Date Opened: 1973
Additional Information: On 13 September 1955, Premier Cahill announced an international competition for the design of a Sydney Opera House. It was built on Bennelong Point between 1959 and 1973, to the award-winning design of the Danish architect Joern Utzon, and opened by Queen Elizabeth II in October 1973.

Palm House (see Glasshouse No. 10) Glasshouses

Palace Garden Pavilion (see Rose Garden Pavilion)

Pergolas

Date built: 1907, 1990
Additional Information: The second pergola in the Gardens (after the one built in the region of the Succulent Garden in 1832 — see p.28) was built to the west of the Herbarium (after the removal of the old glasshouses) on the site of a rose garden. The Italian Pergola, constructed of native timber, formed an avenue terminating at the office door.[60] It was reconstructed at some stage before 1936 with sandstone pillars and moved to the southern side of the Rose Garden Pavilion in 1957.

> The stone-and-wood pergola which formerly led to the entrance of the general office was destroyed by a short, violent storm. It has been re-erected along a path leading to the stone-based shelter pavilion in the Garden Palace Grounds.[61]

This pergola was then incorporated into the design of the new Rose Garden (1987); the matching pergola (paid for by the Friends of the Gardens) was built in 1989/90, and acknowledged at a ceremony on 8 April 1990.

(Governor) Phillip Wall

Date built: 1988
Additional Information: This memorial wall in Bed 68, constructed from bricks taken from the foundation of the seventeenth-century home of Arthur Phillip, Vernals Farm, Lyndhurst, Hampshire — where Phillip lived before his historic voyage to Australia — was unveiled by the Governor, His Excellency, Air Marshall Sir James Rowland, on 28 January 1988. The bricks were carefully dismantled from remnants of the house and some were constructed here in English bond, bedded in lime mortar as they stood in England in 1787.

The bower took the form of a teak seat and the brick wall — made possible by the enthusiasm of Mr Geoffrey Cottee, a Sydney businessman. Long interested in history, Mr Cottee spent several years tracking down the area where Phillip lived from 1763 to 1774. He narrowed the search to four sites listed in contemporary books, then to two. Finally, an old reference to a toll house on the road to Lymington led him to the remains of Vernal's Cottage, a forgotten ruin hidden behind a hedge. Then began the long process of recording, dismantling, transporting and reassembling most of the wall — leaving some of the bricks in-situ. By the end of May 1987, every brick was documented, arrangements were made for their transfer to Sydney, and one symbolic brick was carried by the First Fleet re-enactment.

Mr Cottee's researches took him to the Lyndhurst Park Hotel, formerly Glassheye's Cottage, which stood on land leased by Phillip. In the grounds of the hotel is a piece of Sydney Cove sandstone marking the connection between Lyndhurst and Australia. Another link nearby is the site of Cuffnells, the house built for Sir George Rose, British Secretary of the Treasury in the eighteenth century, who had an undoubted influence on the choice of his near neighbour to lead the First Fleet. Cuffnells was later the home of Alice Liddell, the Alice of Lewis Carroll's *Alice in Wonderland.*

On the same day, an oak tree grown from an acorn from an ancient tree in the original garden of Vernals Farm was planted nearby in Bed 68 (see Phillip Oak, p.65).

Pyramid Glasshouse (see Tropical Glasshouses)

Refreshment Room (see Kiosk)

Restaurant on the Park (Pavilion on the Park)

Date built: *c.* 1964
Additional Information: The Restaurant, built on the site of the old Bandstand, after the construction of the Cahill Expressway. It was damaged by fire in 1989.

Rose Garden Pavilion, (Palace Garden Pavilion/Chess Pavilion)

Date built: 1897
Additional Information: The Garden Palace was burnt to the ground on 22 September 1882, and the grounds were then gradually incorporated into the Gardens.
When Joseph Henry Maiden became Director in 1896, he was anxious to instal a shelter pavilion of the type he had seen when he was in Germany, where people met and played chess.[62] The pavilion, designed by Walter Vernon, Government Architect, was subsequently built in 1897, for the sum of £400 10*s*. 6*d*. and was called the Chess Pavilion, as epitomised by the central 'castle' or 'rook' and the four separate 'pawns' on its roof.[63]

A second, similar structure was built in Centennial Park near Park's Drive, opposite the duck pond, in 1898. This pavilion at one stage had windows which were later removed.

The pavilion was constructed from with massive stone piers of Port Jackson sandstone[64] supporting a wooden latticed frieze, and an elaborate roof structure with decorative timber fretwork to the eaves rafters, and an ornate, central hanging timber boss. The roof, originally of she-oak (Casuarina) shingles, in a swastika plan — a square with radiating bays to the left of each elevation — was replaced with slate early this century.[65] The original wooden slatted seats remain and the building was classified as 'Significant at the National Level' by the National Trust in 1976.

In the preparation of the foundation and floor, the space to be occupied by the building was excavated to the bedrock. The area had been filled in with town rubbish and debris 12 years previously, subsequent to the laying out of the grounds after the burning of the Garden Palace. The unsightly cement foundation which had supported the Queen's Statue in bronze in the centre of the building (see p.120) was removed at the same time and was broken up to form the floor of the pavilion. The adjacent ground was levelled with soil brought in from the new wing of the Art Gallery being constructed in the Domain.[66]

In 1901 the adjacent flight of steps had been constructed of Pyrmont sandstone, and an asphalt path constructed to link these steps, the pavilion and its southern entrance to the circular rose garden that was then at the spot of the centre of the Garden Palace dome.

The pavilion had proved to be very popular and, prior to the construction of the steps, the public had been damaging the embankment by taking a short-cut down the grassed slope. At the same time the surrounding lawn, which had previously sloped in an eastern direction, was raised using a large quantity of soil (some of which came from the site of the government tramway shed, then being erected at Macquarie Point, where the Opera House now stands).[67]

The Tram Shed, Bennelong Point, *RBG Library.*

By 1902 a new garden border was established at the base of the embankment. About 100 loads of debris (town rubbish) were removed in the process and 100 loads of good loam introduced, which was then planted with tea roses and other herbaceous plants and annuals.[68]

By 1907, the pavilion floor was becoming very wet because it had subsided over the years as the surrounding lawn was raised. The original surface was broken up (by Public Works) and a new floor laid down.[69]

In 1919 the roof of the pavilion in Centennial Park had been changed over from wood shingles to slate tiles, the tiles being obtained from the old Zoo (formerly in Moore Park). It was reported that the tiles were well weathered and that these gave the roof a pleasing and soft effect of age.[70] No reference has been obtained relating to the change from shingle to slate of the pavilion in the Gardens. As the two buildings were under the same management, it is probable that this change was also made at the same time. An inspection of the two buildings supports this theory.[71]

The Rose Garden, designed by the Director, Professor Carrick Chambers and officially opened by Lady Rowland on 10 April 1988, has incorporated the pavilion and pergolas (see p.64) as a special feature. Consequently, in 1988, its official title became the Rose Garden Pavilion.

As part of the development of this garden, roses were again planted on the embankment and hanging roses were planted in four Cliveden Vases. The vases were obtained from Haddonstone Australia for a total of $1,140, and mounted on the previously mentioned stairs in the winter of 1987. Early photographs from both 1907 and 1920 show that there were no urns on these stairs at these times,[72] although an undated photograph from the Gardens' archives (calculated at 1901 because of the absence of a garden border at the base of the embankment) shows two urns and two statues on the stairs. Unfortunately two of the later urns were vandalised on the night of 1 November 1989.

Sea Wall

Date built: 1846–78
Additional Information: Work began on the retaining wall — the Sea Wall — around Farm Cove in 1846.

> Men [were] now employed in building a wall all around, so as to extend the esplanade at the water side from one extremity of the garden boundary to the other, leaving an opening for the creek which empties itself below the willow pond, (on which stands Allan Cunningham's Obelisk), over which a bridge will be thrown.[73]

The Sea Wall, 1989, *Jaime Plaza.*

Prior to that the tidal flat of Woccanmagully had extended up to where the Gardens Restaurant now stands. The flat was built up with silt dredged from Sydney Cove and with fill from the construction sites of city buildings. Construction was carried out by the family firm of the botanist Joyce Vickery's mother, Ottomar Rossbach.[74]

It turned out to be a 'Thirty Years' Wall' as it was not completed until 1878. Amongst Moore's troubles, most of the Gardens staff left for the diggings in the gold rush of 1852.[75]

State Library of New South Wales

Date built: 1910-42
Additional Information: The State Library of New South Wales, with its Mitchell and Dixson Libraries, stood partly on the site of the old (Female) School of Industry (founded by Governor Darling in 1826 in the old light-horse barracks for training domestic servants) and partly within the Macquarie Domain. The School of Industry (1826–70) had been established to train domestic servants in 'every branch of household work'. Late in the 19th century, this ground was waste, until the Director of the Gardens, J. H. Maiden, established an acacia plantation there.

The Mitchell Library was built between 1904 and 1910 for a contract price of £23,000 and opened on 8 March 1910 by the Governor, Lord Chelmsford.[76] The greater complex was not completed until 1942. With the completion of the new extension of 1988, the State Library moved its main entrance to the Macquarie Street frontage; the old Reading Room becoming an expanded Mitchell Library.

St Marys Lodge (see Domain Lodge)

Succulent Garden

Date built: *c.* 1983
Additional Information: The Succulent Garden, designed by M. A. Watson for the site of the old aviaries and zoo, was constructed over several years from clinker brick and green slate (in a 1970s 'vertical patio' style) to create a separate microclimate for succulent plants.

Toilets

Date built: 1897 to present
Additional Information: There are no toilets indicated on the 1870 or earlier plans of the Gardens. 'Public latrines' are shown on the 1897 plan, in the location of the existing toilets in the eastern part of the Lower Garden.[77] However, in 1914:

> the accommodation of both the Women's & Men's lavatories, which for years have been inadequate and out of date, has been more than doubled, and the lavatories fitted with the latest sanitary conveniences. This work was the direction of the Government Architect's Department, at a cost of £1,275.[78]

These toilets were located to the south west of the Kiosk and on the site of the original toilets, as shown on the 1914 plan of the Gardens. In 1932 the ladies lavatory near the Kiosk was extended, and a new toilet block for men was constructed adjacent to the Depot gate in 1945. The toilets on the original site were replaced by a new building in the 1970s.

In his evidence to the Royal Commission on the Improvements of the City of Sydney in 1909, J. H. Maiden stated:

> The provision of water-closets and urinals in public parks is a matter of absolute necessity ... We find funds for services which are not as absolutely necessary for healthy existence of the citizens as urinals and water closets. You cannot expect a man to enjoy your park improvements ... if his physical condition points to distress, and even agony, leading to impaired health ... Surely, gentlemen, we cannot have a beautiful Sydney without a happy and a healthy Sydney, and you can have neither the one or the other without attending to the public conveniences, as we somewhat vaguely call them.[79]

The parkland of the Domain was far worse serviced than the Gardens in this respect. Public latrines for men were erected in the Domain

> in 1895, and are now extensively used by the public. The present accommodation is already rather limited, and arrangements are being made for its numbers are not decreasing. The latrines require frequent attention to keep them wholesome and clean, and improved arrangements will be adopted with the extension referred to. These latrines are confined to one portion of the Domain, in a convenient, though not in a central situation. In my opinion more latrines for men, in the vicinity of Mrs Macquaries Chair, are absolutely necessary. The absence of similar accommodation for females is a matter for very great regret.[80]

It was not until 1905 that a Ladies' Reserve was being considered for the Domain:

> While excellent provision for 'men only' has been provided in one corner of the Domain, and it is very largely availed of, no similar consideration has been given to the other sex. For this purpose a small area could be enclosed, and laid out more or less ornamentally, as a 'ladies reserve', absolutely restricted to the sex, and furnished with a few seats and flower beds, &c. Within this reserve lavatories for 'ladies and children only' could be constructed, which I feel certain are justified on the grounds of humanity.[81]

Toilets (for both men and women) were finally erected on the northern peninsula towards Mrs Macquaries Chair in 1910.[82]

Twin Cottages (Residences Cottages 5 & 6)

Date built: Completed 1899
Additional Information: The semi-detached two storey red brick cottages, with Marseilles tiled roofs and terracotta cresting and casement windows, were designed by Walter Vernon, Government Architect. Strongly influenced by the English arts and crafts movement, they were built by Foster Brothers for £1,260 (£68 18*s.* 6*d.* for fences and gates):

> Two cottages ... have been erected for the indoor propagator and the gardener (hot houses). The site (overlooking Woolloomooloo Bay) is commanding one, and

Victoria Lodge looking across towards the Conservatorium of Music, after 1864. The Victoria Lodge Gate (1864) was moved to become Lion Gate in the Macquarie Wall in 1901. *Government Printing Office.*

> enables these officers to exercise some supervision out of hours over both Gardens and Outer Domain, while the cottages, which are very picturesque, are an ornament to the landscape.[83]

Victoria Lodge (Cottage 3)

Date built: 1863 or 1865
Additional Information: The Statistical Register of New South Wales[84] lists the construction of two lodges in this period — a lodge completed in 1863 at a cost of £300, and the Eastern Lodge, completed in 1865 at a cost of £476 11*s*. 7*d*. Victoria Lodge is indicated but not identified on the 1867 plan of the Gardens, but is identified as Victoria Lodge on the 1870 plan, while Eastern Lodge is not mentioned.

Charles Moore, in his 1870/71 Report to the Legislative Assembly[85] refers to the 'relaying of the upper part of the [Lower] Garden near the Eastern Lodge'. Arguments of symmetry would suggest that this was Victoria Lodge, as there was in fact a Western Lodge in the Gardens (built in 1856 of brick, and demolished in 1912/13, see p.22).

The stone piers of the Victoria Lodge Gate (marked 'VR' Victoria Regina in the Macquarie Wall near Lion Gate Lodge — moved there in 1902) are dated 1864. It is logical to assume they would have been installed when the Lodge was being built.

The small, single storey asymmetrical Italianate Lodge with a two-storey tower was probably designed by the Colonial Architect, James Barnet. Now the house of the Gardens' Ranger, it was built of stone, with stone mullioned windows, in stages, over a number of years. A photograph published in *Discovering the Domain* shows Victoria Lodge without the southern wing with the present Conservatorium of Music in the background.[86]

Vista Pavilion (Octagonal Shelter Pavilion)

Date built: 1907
Additional Information: The octagonal shelter pavilion in Lawn 54, designed by Walter Vernon, Government Architect, was built in 1907, at a cost of £184. It was built to replace the rustic summer-house which stood in the Arrangement Ground for over 40 years, being finally removed in 1902.[87]

Chronological Checklist of Exisiting Buildings, Walls, Pavilions, etc. in the Gardens and Domain

1812	Macquarie Wall
1821	Conservatorium of Music (Stables)
1835	Domain Lodge (St Marys Lodge)
1840s	Government House Lodge
1845	(new) Government House
1846–78	Sea Wall
1863 (?)	Cottage 4
1863/5	Victoria Lodge
1870+	Bush House (Fernery)
1874	Cunningham Building
1876	Glasshouse No. 10 (Palm House, Education Glasshouse)
1878	Anderson Building (second stage 1899)
1878	Lion Gate Lodge (Cottage 2, Middle Lodge)
1879	Stables (Depot)
1886	Kiosk (see Gardens Restaurant)
1896	Art Gallery of New South Wales commenced
1897	First Toilets in Gardens
1897	Rose Garden Pavilion (Chess/Palace Garden Pavilion)
1898	Packing/Potting Sheds/Storeroom/Kitchen (Depot)
1899	Glasshouse No. 7
1899	(old) Herbarium
1899	Twin Cottages
1900	Glasshouse No. 3 (Fern House)
1907	Vista Pavilion (Octagonal Shelter Pavilion)
1907	Pergola (now southern side of Rose Garden Pavilion)
1908	Fleet Steps
1909	Glasshouses Nos. 5, 6 and 9
1910	Mitchell Library
1911	Glasshouse No. 4 (Show Pavilion)
1921	Cryptogam House (Aquatic House)
1923	Glasshouse No. 8
1929	Maiden Memorial Pavilion
1942	State Library of New South Wales complex completed
1961	Glasshouse
c. 1964	Restaurant on the Park
1965	Glasshouse
1973	Opera House
1976	The Pyramid Glasshouse
1978	Gardens Restaurant (incorporating wall of old Kiosk)
1982	Robert Brown Building (new Herbarium)
1988	(Governor) Phillip Wall
1988	Gardens Palace Kiosk
1989	Pergola (northern side of Rose Garden Pavilion)
1990	(The) Sydney Tropical Centre — the Arc and the Pyramid
1992	(new) Fernery

5

Arcadia Revisited

Plato began teaching in the tree-planted Academy, thus creating the long-held association between philosophy and gardens, revived by the humanists of the Renaissance. One of the first botanical gardens in the Athenian Lyceum displayed plants brought back from the campaigns of Alexander the Great. Here Aristotle's followers had paced their walks, or peripatoi, and thus became known as the Peripatetics. The Romans subsequently adorned their country villas with game parks, aviaries and ornamental fish ponds. Surban villas of the wealthy featured axially arranged gardens with a central water course, surrounded by symmetrical paths and flower beds.[1]

The presence of many Italian statues in these Gardens became apparent to me during the course of this work. Yet more than anything else, the Sydney Gardens reflect the English landscape movement of the eighteenth century — when statues, urns, and grottos were used in association with ruins as focal points of vistas, and as things of beauty in their own right.

Statues were an important part of parks and gardens of the Victorian era. Albert the Good — consort of Queen Victoria — was very keen on the so-called uplifting of the lower orders. Thus those who could not afford a 'grand tour' were to be given a de facto classical education — by the incorporation of copies of archetypal statues into public gardens — where people could stroll in a pleasant environment and learn at the same time. It later became fashionable to make fun of Albert for, apart from being a German, he was an intellectual — but he was absolutely right about the educational value of statues in public gardens.

For gardens, as well as being places of beauty and quiet contemplation, are places for walking and talking — about gardens, or philosophy, or love, or art, or office politics . . . or even gossip. And in this respect the statues may act as beacons along the way — as talking-points in themselves, or as points of geographic recognition — and triggers of associations. And given that some of our statues have been standing in the one place for more than 100 years, they have been visited by generations, as focal points of learning, and happy memories.

Kathleen Wall told me that her parents used to bring her to the Gardens when she was a little girl.[2] Each time they would ritualistically stop at the statues of The Boxers and her father would tell her all manner of pugilistic stories, and then they would move on to the operatic statues (Linda and the Sweep Boy). Beside the Sweep Boy her father told her stories of such Dickensian horrors that she burst into tears, at which point her mother said, 'Stop terrifying the child so'. The next time they came to the Gardens, young Kathleen, just begging to be terrified, took her father by the hand and led him straight to the Sweep Boy, urging him to tell her the story all over again.

In my job I have frequent requests from people regarding the whereabouts of certain statues — dimly remembered from youth or childhood — while others have sharper memories, like Kathleen Wall. In 1971 Mr H. Dominy of Parramatta wrote wistfully of the demise of Rebecca at the Well — one of the group of statues imported from Italy in 1883:

> I experienced a strange sadness when I no longer see a particular object which was so familiar a sight until a few years ago. I refer to the statue of Rebecca at the Well. I never ceased to admire its beauty and simplicity [and] it typified an ideal which could be related to our everyday life . . . with faith, love and courage. To me at least and doubtless to many others this was what the statue represented and it is a shame not to see it there any longer.[3]

The dreadful and prosaic truth was that Rebecca had been 'irretrievably damaged by a contractor's truck'.[4]

Statues in Palace Garden, *RBG Library.*

Linda da Chamonix and the Sweep Boy in Palace Garden *c.* 1900, *RBG Library.*

Many of the statues people seek, like Rebecca, have long since disappeared. Some of them even stood in other parks or gardens, and many have been vandalised. Yet, almost miraculously, a few of them remain, and that only adds to their fascination and their charm.

This chapter is a list and celebration of the surviving statues, etc. in the Gardens to date, with the Domain being considered in Chapter 6.

Alphabetical Checklist of Existing Statues, Fountains and Memorials in the Gardens

Banks Bust

Location: Atrium of the National Herbarium of New South Wales (Robert Brown Building)
The commission of this bust was arranged by the science historian, Anne Moyal, in association with the Royal Society, London, and the Australian Academy of Science. It was sculpted by Mrs Ninon Geier, an artist at CSIRO who lives in Canberra, and whose other works include the most distinguished bust of Manning Clark. The bronze head, cast in the Sydney foundry (Austral Brass Foundry Pty Ltd of Redfern), is one of two prepared from the original sculpture; the other is located in the National Botanic Gardens Canberra. The head in the Royal Botanic Gardens Sydney was unveiled by Sir David Smith, FRS, Vice-Chancellor, Edinburgh University, at a function attended by the artist and approximately 60 guests and staff on the evening of 2 September 1988. Sir David is an authority on Banks and was on an Australian Bicentennial lecture tour of all states.

Bird Bath and Fountain

Location: Central Pond, Lower Garden
The bird bath and fountain installed some time prior to 1892, probably in 1879 leading up to the International Exhibition, had been previously damaged. A new bowl of Carrara marble was obtained from Italy through Melocco Stonemasons for $5,000 and installed in January 1987.

Bird Bath (Sydney Garden Club)

Location: Bed 130, near Opera House Gate
In 1981 the Sydney Garden Club donated a bird bath to the Gardens in the form of a naturalistic rock with a depression carved in the top to catch rainwater. It was installed in the garden near the Opera House Gate. The plaque reads: 'Donated to the Royal Botanic Gardens by the Sydney Garden Club 1981'.

Over a hundred species of birds have been seen in the Gardens over a period of time. On any given day a birdwatcher would probably only see about 30–40 of these. A leaflet on the birds in the Gardens may be obtained from the Visitor Centre. About 13 species of birds seen in these Gardens have been introduced into this country.

Boy Extracting Thorn

Location: Bed 98
This is a copy of the well-known bronze of the Spinario or Boy with a Thorn in his Foot, in the Capitoline Museum in Rome, itself a Roman copy (perhaps of the first century BC) of a Greek original by an unknown sculptor in the third century.[5] Copied many times this life-sized, naturalistic statue has possibly become the most famous 'domestic' sculpture in the world. This particular copy was one of the consignment of eight marbles imported from Italy in 1883. Old photographs establish that it was in the same position in 1903. Unfortunately over the years the statue has been considerably vandalised.

The Boxers

Location: Bed 79 and opposite on Lawn 15, Palace Garden
Two-metre high (approximate) white marble copies of the Boxers, Kreugas and Damoxenes, by Antonio Canova (1757–1822) on sandstone plinths, pugnaciously facing each other across the pathway, and expressing the artist's

Boy Extracting Thorn and period style rockery, *RBG Library*.

The Boxer Kreugas in Palace Garden, 1989, *Jaime Plaza*.

conception of 'energy'. The story of The Boxers was first told by the Greek historian Pausanius about 300 years before Christ. These two fighters were famous in Ancient Greece and a contest was arranged. They agreed that each must in turn let himself be hit by his adversary in whatever part of the body nominated. Kreugas won the right to have first blow and struck Damoxenes on the head. Damoxenes was not stunned and then ordered Kreugas to hold his arms up high. Damoxenes hit Kreugas in the stomach with his fingers extended and tore out his innards. Our statue shows Damoxenes about to make the fatal blow.[6]

In 1800 Napoleon Bonaparte's army conquered the Italian states and took some of the best classical art treasures back to France. To replace some of the lost statues the Pope then commissioned Antonio Canova, the greatest sculptor since Michelangelo, to make two Carrara marble statues of Damoxenes and Kreugas. In 1802 the statues were placed in the Belvedere Court of the Vatican and are there today. The statues were very popular and Canova's workshop produced some copies. Other copies were made over the years.

Old photographs confirm that the Boxers have been in the Botanic Gardens for at least 100 years. Their source is unknown at this stage. They did not come from the Art Gallery of New South Wales, and they are probably copies from Canova's studio, presented to the Gardens by Sir James R. Fairfax.[7]

On the evening of 12/13 December 1990 they were attacked with a star picket, and Kreugas lost the index finger of his right hand.

The Boxers were also to be found in Centennial Park (imported in 1888), but were vandalised during the early 1930s when large numbers of unemployed men were living in the Park. To prevent further damage they were stored until about 1974. One of the surviving Boxers, Damoxenes, was given to Hurstville Boys' High School in 1976. He was still there in 1988.[8]

Choragic Monument of Lysicrates

Location: Lawn 39, Lower Garden

This replica of the Choragic Monument of Lysicrates was executed by Walter McGill, sculptor, for Sir James Martin, Premier of New South Wales, and was erected in the grounds of his residence, Charlemont (then known as Clarens) in Macleay Street, Potts Point, in 1870.

The cylindrical and solid monument, 5 feet in

diameter and 15 feet high, was carved from local sandstone, working from measurements taken from the original by Stuart and Revett in 1750, as published in 1824. Walter McGill, the sculptor, lived in Victoria Street, Darlinghurst, and is believed to have also designed and carved the piers to the gates at the entrance to the Gardens near the Art Gallery (see Woolloomooloo Gate, p.38). He also carved a figure of Captain Cook at Randwick, and worked on the Town Hall and General Post Office, Martin Place.

Sir James Martin, after whom Martin Place was named, was the son of an Irish groom who had accompanied Sir Thomas Brisbane to Australia and remained here. The son James, born in Ireland in 1820, was educated at Timothy Cape's School and the Sydney Academy and became a classics scholar. After completing his education in 1836, he became a journalist and later a lawyer, and was elected to the Legislative Council in 1848 at the age of 27. He supported Wentworth's scheme for the University of Sydney and Lamb's on the abolition of transportation, and was Premier of New South Wales in 1863, 1866 and 1870. He was knighted in 1866, became Chief Justice in 1873, and died in 1886.

When the site of Clarens was acquired during the war by the Commonwealth government, in association with the construction of the Garden Island graving docks for the RAN, the monument was made available to the State, and was removed to the Botanic Gardens at the request of the then Minister for Education, the Hon. Clive R. Evatt. A ceremony to commemorate the move was performed on the site by the Hon. W. J. McKell, Premier and Treasurer, on 16 November 1943.

The original Choragic Monument,[9] now in partial ruin, was incorporated into the grounds of the Monastery of the Capuchins in Athens in the eighteenth century, close to the Theatre of Dionysus. It was first erected in 334 BC, some four years after the Battle of Chaeronea, in which the city states had been conquered by the Macedonians.

The significance of the monument was that it marked the continued practice of one of the important Athenian ceremonies — like a cross between a sports day and an eisteddfod — conducted by the richest of the citizens at their own expense. Part of the custom involved this citizen-patron or 'Choragus' (hence the term 'choragic') being responsible for the training of a particular team (or tribe) in the athletic, musical or dramatic activities at the Dionysia, or Festival of Bacchus, each year.

Choragic Monument in original location at residence of Sir James Martin, *RBG Library*.

The purpose of the monument was to display the winner's prize — a brass tripod received by the sponsor choragus on behalf of the winning team or choregos — the catch being that the 'Choragus' was then responsible for the expenses of its consecration to Dionysus and the construction of the sacred edifice (monument) on which it was placed. The tripod itself was fashioned on an earlier tripod — a particular seat where the Delphic priestesses sat when they pronounced the oracles and inspired sayings. This class of monument once lined a street in Athens — known as the Street of Tripods.

This specific monument, the Choragic Monument of Lysicrates, was built to receive the victor's tripod won by Lysicrates of Kikyna (as inscribed on the architrave) who had won the drama contest at the Festival of Bacchus in that year (334/3 BC) with his chorus of boys. The inscription on the architrave also records that the tribe of Akamantis won the victory: Theon played the flute; Lysiades of Athens trained the chorus; and Evainetos was Archon (Chief Magistrate) of the year in which the contest took place.[10]

The Choragic Monument is of special interest to

Greek scholars and students of architecture, and Sydney is lucky to have such a fine model.

The original monument, 7 feet in diameter and 18 feet high, was built of Pentelic and other marbles which gave it a mellow glow, and had three distinct parts — a rusticated base (or stylobate) of Piraeus stone, 9 feet 6 inches square, supporting a closed circular colonnade and a cupola (small rounded ceiling dome) with a tall, central three-sided ornament to take the feet of the brass tripod. The colonnade was constructed of six Corinthian columns, each joined with a single slab of marble, creating a hollow space inside, to which there was no access. At the top of each panel, between the capitals of the columns, two tripods were sculptured in relief. The capitals themselves are among the few early examples of the Corinthian order — with the usual decorative motifs of acanthus (*Acanthus spinosus*).

A fifth-century Athenian sculptor, Kallimachos, had seen an acanthus growing over a basket that had been set on the grave of a Corinthian girl by her devoted nurse. Struck by the beauty of the curling leaves, he created the column capital that is still called Corinthian.[11]

In this case the columns also contain lotus leaves — and are treated more individually than the more 'orthodox' capitals of later times. Inside, where the capitals could not be seen, they were left unfinished.

It is interesting here to note that William Guilfoyle, renowned landscaping director of the Melbourne Botanic Gardens, when designing his Temple of the Winds, used staghorn ferns to top his columns instead of the more traditional acanthus, creating an indigenous architectural form — Australian Classic. Guilfoyle had lived for a time as a child on his father's nursery and experimental farm in the Cudgen scrub in far north-eastern New South Wales. He would have seen the staghorn ferns, *Platycerium superbum*, growing in the local rainforests when all the Tweed Valley looked like Stott's island.

The Temple of the Winds

Young William on his father's plot
in Cudgen scrub, discovering,
embraced the *Platycerium* —
so when the Baron got the chop,
(Germanic order on the block)
he built his temple to the winds —
capped its columns with the things,
not *Acanthus mollis* topped.

Man of Corinth, fashion men
and mollify the savages
to know Australian Classic, when
supporters of some Melbourne club,
cavorting in a Gold Coast pub —
still call the staghorns cabbages![12]

Above the capitals of the Choragic Monument is an entablature and continuous frieze, surmounted by an ornate crested cornice, and crowned by a carved and ornamented marble dome or cupola, imitating a thatch of laurel leaves. Three scrolls branch out from the upper part of the dome, generally supposed to have supported dolphins or figures; the cornice has its peculiar honeysuckle scroll. At the apex, the central foliage stalk of acanthus leaves develops into the three arms on which the votive tripod stood. The frieze, barely more than 10 inches in height, illustrates in bas-relief the myth of Dionysus and the Tyrrhenian Pirates as told by Ovid in his *Metamorphoses*, Book 3.

Dionysus was a native god of fruitfulness and vegetation, especially the grape vine, and thus the god of wine — later to be equated with Bacchus by the Romans. The god, attacked and captured by Tyrrhenian Pirates on a voyage to Naxos, transformed them into dolphins. In this version the sculptor has set the scene on a rocky shore and brought his attendant satyrs to the help of the god. In the centre, Dionysus sits on a rock fondling a panther. On each side are satyrs, attendant on Dionysus, or watching the fight, or rushing to join in, or actually engaging with the pirates. One of the pirates is leaping into the sea, and has already been transformed into a dolphin as far as his waist.

By means of a close but not exact correspondence between the groups to the left and right of the central figure, the sculptor has produced a formal balance with enough variety to avoid monotony. The flying panther skins of the satyrs add to the impression of action and spaciousness of design — and cost Mr Lysicrates, no doubt, a tidy sum of drachma.

The original monument (as measured by Stuart and Revett in 1750) was defaced by fire during the war with the Turks in 1821, and has considerably deteriorated since casts (now in the British Museum) were made by Lord Elgin.

In 1982 the Choragic Monument in the Sydney Gardens was used as a focal point for meetings of the Australian Byron Society. It transpires that the poet,

George Gordon, Lord Byron, born in the year of European settlement of Australia, had an interesting association with the original Choragic Monument in Athens.[13] In 1804 Byron was staying at the Capuchin Convent in Athens while the monks taught him Italian and Latin. He had a study inside the hollow Monument and wrote part of *Childe Harold* there in 1810–11.[14] Apparently a door had earlier been installed and the circular recess of the upper storey was his favourite retreat, separated from the rest of the convent by a green curtain.

In 1765 the Monument had been used as a dormitory for a Greek servant of the monks. Later it was made into a small library, capable of holding just one student at a desk. When Byron arrived it was being used by a boarder called Giuseppe, but it was apparently then restored to its former use as a library for the poet.

The convent stood on the site of a small public garden, majestically overshadowed by the east face of the Acropolis where Padre Paolo introduced Byron to the Greek 'underground', in the narrow streets of the old city. The monks had incorrectly called the Monument the Lantern of Demosthenes — like the Martello tower of Stephen Dedalus (James Joyce) crossed with a London post box — in the mistaken belief that the ancient orator had prepared his political thunders within its hollow, drum-like walls.[15]

There is, however, no doubt that Byron worked in the Choragic Monument. He died of fever in Greece some 20 years later, aged 36.

The convent, unfortunately, was burnt down during the War of Independence, but the Monument was saved and still stands in Vironos (Byron) Street, in the old part of Athens, between the Arch of Hadrian and the east face of the Acropolis.

Byron was to leave his mark on Greece in at least one other way, in the form of his name carved into one of the columns of the Temple of Poseidon (the Sea God) at Cape Sunion, on the coast, half an hour's drive from Athens.

Cometh up as a Flower

Location: Previously in the southern end of Bed 17 near the Visitor Centre — relocated to garden near the Cunningham Building, *c.* 1982

1,200 mm high white marble statue of female, 1878, by the British sculptor, Shakespeare Wood (*d.* 1886), on a square sandstone plinth. This statue, a gift of Mr E. R. Fairfax, 1889,[16] was transferred to the Gardens by the Art Gallery of New South Wales in 1958. 'Cometh up as a flower' is a Biblical reference from Job 14.1,2: 'Man that is born of woman is of few days, and full of trouble. He cometh like a flower, and is cut down; he fleeth also as a shadow, and continueth not.'

Cometh up as a Flower, 1989, *Jaime Plaza.*

Cunningham Memorial

Location: Obelisk in pond on north-eastern side of Gardens Restaurant. His headstone has been incorporated in the external wall of the Herbarium (Robert Brown Building) near the entrance.

Allan Cunningham, explorer and botanist, was for a short time Superintendent (Director) of these Gardens.

In 1831, after the death of Charles Fraser, the first Superintendent of the Sydney Gardens, the position was offered to Allan Cunningham. He declined, and the job was accepted by his younger brother Richard who arrived in Sydney in 1833, to be clubbed to death by natives on Thomas Mitchell's expedition down the Bogan River in western New South Wales in 1835. A few fragments of the remains of Richard Cunningham were collected by a Lieutenant Zouch of the Mounted Police, and buried at Lower Tabratong, near Dandaloo, where a stone marks his last resting place (on what was the Hunt Brothers' Burdenda station).[17]

There is a memorial to Richard Cunningham in Saint Andrew's Scots Church Rose Bay 'erected [as an] affectionate tribute to his memory by Allan, his only brother. The Memorial contains the following inscription:

> Richard Cunningham
>
> Government Botanist of this Colony attached to an explorative expedition into the Interior under command of Major Mitchell, Surveyor General, wandered in his enthusiasm for botanical investigations from his companions and losing himself in the desert country on the Bogan River fell into the hands of one of the native tribes by whom he was unfortunately killed [in the] 42nd year of his life.

St Andrew's Scots Church was originally opened on 13 September 1835, in Kent Street, Sydney, behind where St Andrew's Anglican Cathedral now stands. The church was closed in 1911 and moved to the corner of Dover Road and Carlisle Street, Rose Bay, in 1913.

After the death of his brother, Allan accepted the job of Superintendent. Born in London in 1791 Allan Cunningham became a plant collector for Kew and a protegé of Sir Joseph Banks. He was sent to Brazil in 1814, and arrived in Sydney in 1816.

He explored much of south-eastern and north-western Australia between 1817 and 1828 and collected many plant specimens — mostly leaving their description and publication to others. He also found a convenient access to the Liverpool Plains, discovered the Darling Downs and the pass linking them to the coast, to become, with the exception of Robert Brown, the most accomplished of botanists to visit Australia in that period, and certainly one of our most successful explorers.

In 1831 he returned to England, but he returned to Australia in 1837, from 12 February, as Colonial Botanist and Superintendent of the Sydney Gardens. He was only 45 years old but was debilitated by the hardships of years of rough travel in the bush.

His second sojourn in Australia was not a happy one. A Committee of Management had been set up and had taken over the institution he was supposed to control. There was a considerable dispute over the demands of 'all-devouring officials' for fruit and vegetables grown in the Gardens. Cunningham was essentially unhappy at having to supervise the kitchen garden of Government House and maintain 'pleasure grounds for the citizens of Sydney', which left him little time to devote to botanical research. He resigned from the position of Superintendent of the Gardens in January 1838. The paper ran a story on the matter which makes interesting reading today:

> The Botanical, alias the Kitchen Garden —
>
> We have had frequently to call the attention of the Colonists to the fact, that a kitchen garden, under the pretence of being a Botanic Garden, is supported in Sydney at an expense of from Eight Hundred Pounds to One Thousand Pounds a year . . .
>
> We scarcely ever walk through this garden without seeing some servant, with a basket, carrying off vegetables or fruit for Mrs This, or Mrs That — the wife of some official. Can't these people go to market and purchase their supplies as independent persons do, instead of poaching on what is really public property? . . .
>
> Seriously we do say, that such an impudent job ought to be done away with. It is, in fact so barefaced, that Mr Cunningham would no longer consent to remain a mere cultivator of official cabbages and turneps [*sic*], and accordingly he has resigned the management of the Botanic Garden in disgust.[18]

The Governor promptly ordered that 'no culinary vegetables of any sort shall be raised in the garden and that the fruit-bearing trees shall gradually be removed'.

By 1846 'the orange and other trees . . . [were] already fast failing'.[19] Some eight olive trees remained and Mr Kidd (the Superintendent) succeeded in producing an oil 'which has astonished, by its colour, its flavour, in fact in every way'.[20]

Cunningham, his health declining, sailed for New Zealand to collect specimens on his own account, collected a 'chill' and returned to a cottage in the Gardens 'in a deplorable state of health. He died there on 27 June 1839, aged 48, in the presence of James Anderson, the next Superintendent of the Gardens. Captain Phillip Parker King eulogized: 'He was a rare specimen — quite a genus to himself; an enthusiast in Australian geography; devoted to his own science, Botany; a warm friend, and an honest man'.

Allan Cunningham was buried in the old Devonshire Street Cemetery, where Central Railway Station now stands. Five years later, in 1844, an obelisk was erected to his memory by some of his admirers on a tiny island in the top pond of the Lower Garden (before the erection of the wall around Farm Cove, when the tide used to

Cunningham's Obelisk, with willows grown from cuttings from the willow that grew beside Napoleon's tomb at St Helena, *RBG Library*.

come up to the monument). The inscription on the obelisk read: 'Erected to the memory of Allan Cunningham, Botanist, MDCCCXLIV'.

In 1901, when construction began on Sydney's Central Railway Station, Allan Cunningham's brick grave was opened. His remains (consisting of a piece of skull about four inches in diameter, a piece of large bone from the arm, and some other fragments) were removed and placed in a small leaden casket, 10 inches long, 5 inches broad and 4 inches deep. The casket was placed in a hole that had been made in the obelisk on 26 June 1901, and covered with a marble slab, with the following inscription in leaden lettering: 'The remains of Allan Cunningham were interred in the Devonshire Street Cemetery in July, 1839, from which they were reverently removed on 25 May, 1901, and placed within the obelisk'.[21]

These days the obelisk is in the pond north-east of the Gardens Restaurant. On the island behind is a grove of Bangalow palms, *Archontophoenix cunninghamiana*, one of the half-dozen Australian trees with which fellow botanists have honoured his name.

There are several other monuments to Allan Cunningham around Sydney.

In 1902 a tablet was also erected in St Andrew's Scots Church, Sydney (now at Rose Bay), beside the one he had earlier erected to the memory of his brother. The inscription in part reads:

> Allan Cunningham . . . associated in the pursuit of botanical discovery, with Oxley, in exploring the interior of New Holland, with King, in four times circumnavigating its coasts, and by subsequent personal research, having more fully developed the geography, and flora, of the northern districts of this Colony, and of Norfolk Island, and New Zealand. He has left enduring monuments of devotion to the cause of science and of eminence in those branches which he most assiduously cultivated. Frank, unaffected, firm in principle, with warm feelings, tempered by a most kind and benevolent heart, deservedly beloved by his friends, some of them in the foremost rank of science in England, France and Germany.

The Department of Lands building also has a statue of Allan Cunningham at the corners of Bent and Loftus Streets.

Most recently, when the National Herbarium (Robert Brown Building) was being constructed (*c.* 1982) his original tombstone was incorporated into the external wall near the main entrance off Mrs Macquarie's Road which reads: 'Sacred/to the memory of/Allan Cunningham Esq./Departed this life on the 27th June 1839/Aged 48 years'.

J. H. Maiden, writing of Allan Cunningham, said he would 'shed lustre on the Sydney Botanic Gardens. He is the only man whose statue . . . can appropriately be placed [in the Gardens]'.[22]

Darnell-Smith Memorial Seat

Location: Bed 83, on bank behind Conservatorium of Music

George Percy Darnell-Smith, DSc, FIC, plant pathologist, was Director of the Gardens from 1924 to 1933. When he died in 1942 he bequeathed the sum of £100 to the Gardens for the erection of a 'stone seat overlooking Sydney Harbour'. The Trustees of the Gardens decided to regard the seat as a memorial to Darnell-Smith.

Nothing however was done during the Depression and war years, and then there was a delay over site considerations. A position which the Trustees favoured was at that stage under consideration for a memorial to King George V.

By the time the Trustees had selected the present site the funds were insufficient to build a fitting memorial. The solution was to incorporate a memorial seat into a beautification scheme that was being carried out in that area. The seat, in a secluded sandstone brick enclosure, now covered with the exotic weeping fig, *Ficus pumila*, was completed on 6 June 1952. The enclosure remains but the seat has been removed (1991).

Dolphin

Location: Twin Ponds (northern end)

A 'cold-moulded bronze' (fibreglass impregnated with bronze) statue of a dolphin, 150 × 40 × 55 centimetres, by the Queensland sculptor Robert Lees, was donated to the Gardens by Robert Donato of Robert Donato Nurseries Pty Ltd in 1988. It was placed in the northern end of the Twin Ponds.

First Farm Memorial

Location: Lawn 5, Middle Garden

J. H. Maiden, Director, Royal Botanic Gardens, 1896–1924, first made the point that, 'In no other part of the world can we point to a spot and say — here is the site of the beginning of the horticulture and

Sunday Afternoon in the Palace Garden *c.* 1910, *RBG Library.*

agriculture of a continent'. From the earliest days the Sydney Botanic Gardens have been associated with the development of agriculture and horticulture in Australia — from the first farm to the present — and in no other continent can such a claim be made.

Legend also has it that the oblong beds in the Middle Garden follow the direction of the first furrows ploughed in Australia. A close inspection of Phillip's map suggests that the furrows ran essentially east-west, as do the dominant pathways today.

The First Farm Memorial, in the form of a bronze plaque, was set in a sandstone base made from the stone left after the relocation of the Main Entrance Gate (now Woolloomooloo Gate) when the Cahill Expressway was being built.

The suggestion that the first farm should be recorded in this manner was made by Dr F. T. Bowman, Principal Research Officer, Division of Horticulture, Department of Agriculture, and was unveiled on 30 September 1952. The plaque reads: 'In these Gardens began the Agriculture and Horticulture of a Continent. The first farm in Australia was established on this site with seeds and plants brought by the First Fleet in 1788.'

When Arthur Phillip arrived in Port Jackson on 26 January 1788 to set up his penal colony, he selected Sydney Cove, with its deep anchorage close to shore for its good source of water, the Tank Stream. His next highest priority was to commence agriculture to provide his charges with fresh fruits, grain and vegetables.

Phillip had brought many plants and seeds with him, including coffee, cocoa, cotton, bananas, oranges, lemon, guava, tamarind, eugenia or pomme roses, apple, and prickly pear and cochineal grubs; castor oil; and ipecacuanha and jalup (medicines) from Rio de Janeiro; and figs, bamboo, Spanish reed, oaks and myrtles; sugar cane, quince, apple, pear and strawberry; plus rice, wheat, barley, Indian corn (maize) and other grains from the Cape of Good Hope.

Early maps indicate that there was a kitchen garden about where the AMP complex is today, between the site of Old Government House and the water. The officers also had their own garden plots. The site for the Government Farm was chosen to the east of the little creek that still flows through the Gardens, because of the presence of some alluvial soil, as the name 'Farm Cove' commemorates today. Phillip wrote, 'The principal farm is situated in the next cove to the east of the town, and less than half a mile from it'.

Clearing the land proved to be a most arduous task, given the condition of both the tools and the men. The removal of tree stumps created even greater difficulties. On 29 May 1788 Watkin Tench wrote, 'It will scarcely be credited when I declare that I have known twelve men employed for five days in grubbing out one tree'. Public Service gardening indeed! By July 1788 Governor Phillip spoke of 'a farm nine acres in corn [wheat or barley]', and in September 1788 Phillip records, 'six acres of wheat eight of barley and six of other grain'.

The Farm, as indicated on early maps, occupied the area of what we now call the Middle Garden, extending from the creek to the boundary of the Garden on the east and down to the harbour inlet at the site of the Gardens Restaurant.

Watkin Tench also reported in 1788 that: 'the Governor had established a government farm; at the head of which a competent person of his own household was placed, with convicts to work under him'. That person was Henry Edward Dodd, valet to Governor Phillip, who had worked as an agricultural labourer on Phillip's property, Vernals Farm, Lyndhurst in Hampshire, and was by all accounts a good choice.

The Irish convict, James Ruse, famous for his successful farm at Rose Hill, Parramatta, may well have worked under Dodd. For Ruse, before he died, had arranged for the following inscription for his headstone: 'and when I arrived in this colony I sowed the first grain'.

The farm, however, because of the poverty of the soil, had virtually folded up within six months, and was moved to the Crescent (Parramatta). On 1 October 1788 a gang of 100 convicts were sent to the head of the harbour (Parramatta) to put 'land into cultivation'. It did not help matters that the convicts were mostly urban, with no experience of farming or gardening.

Lieutenant Governor Ross lamented in 1788 that: 'this country will now answer to settle in, for although I think corn will grow here, yet I am convinced that if ever it is able to maintain the people here it cannot be in less time than probably a hundred years hence'. Ross, the thorn in Phillip's side, was to be proved wrong. On 30 October 1788, Phillip argued: 'If fifty farmers . . . were sent out with their families they would do more in one year in rendering this colony independent of the mother country, as to provisions, than a thousand convicts'.

The Farm Cove establishment was probably still used for vegetable cultivation under the management of

Nicholas Devine, Superintendent of Convicts, who subsequently obtained a 'permissive occupancy of the Governor's Old Farm', and a lease of part of the same in 1794. There is a notice in the *Sydney Gazette* in 1803 regarding a theft of 'apple-tree plants' from this early horticulturist.[23]

After Phillip's departure the Government Farm area was leased out for farming, and at least eight leases (for terms of 10 to 21 years) were granted between 1794 and 1806.

Huntsman and Dogs Statue

Location: Lawn 16, near Rose Garden Gate (previously called Conservatorium Gate) near site of Boston's Mill (see p.16).

This statue, by the French sculptor A. (Henri Alfred Marie) Jacquemart (1824–96) at the Rose Garden Gate, has been well-known to generations. An old photograph of the destroyed Garden Palace indicates that the statue was in the Gardens prior to the fire of 1882, after which it was relocated. 'Huntsman and Dogs' is 'imitation bronze' — an iron casting on which copper has been deposited by electrolysis. At the time of writing it had several rust patches and required conservation.

Huntsman and Dogs was cast by the Val d'Osne Company and two copies were sold to Australia — one to the Sydney Gardens, the other to a family living in Melbourne at the corner of Burke Road and Myrniong Grove, Camberwell, which became a showpiece for tourists long before World War I. Their daughter married Mr J. F. S. Allan, and the sculpture was transported to its present site at 330 Canterbury Road, at the corner of Wantirna Road, Ringwood. The Melbourne sculpture was at one stage exhibited in the sculpture court at the Nicholson Street entrance to the Exhibition Building, as photographed by Nicholas Caire in 1881.[24]

I Wish

Location: Intersection of the two main pathways in the Middle Garden at the site of the original Wishing Tree

Pink concrete head of a girl, 620 millimetres high, mounted on a sandstone plinth bearing the words 'I Wish', presented to the Gardens by Leo Buring, wine merchant of Sydney, and his wife Ida in 1946.

The Czechoslovakian-born sculptor Arthur Fleischmann (1896–1990), who arrived in Australia in 1938, was commissioned to do the work which stands on the site of the original Wishing Tree. (See p.49). Other commissions in Australia included the bronze 'explorers' doors of the State Library of New South Wales and the DNA sculpture in the foyer of the State Library, obtained some 40 years later.

Arthur Fleischmann studied medicine in Budapest and Prague before enrolling in the Art Academy of Vienna in 1921. He left Australia in 1948 and was based in London for the rest of his working life, and is the only known artist to have sculpted four Popes from life. Arthur Fleischmann died in Teneriffe in 1990, aged 93.

Japanese Lanterns

Location: Bed 88

By 1921 a Japanese Garden had been established in what is now Bed 88. Planting of bamboos, etc. had started in 1917[25] and a pergola and two toriis (gateway of Shinto shrine, with two uprights and two cross-pieces) were constructed at the north-western end of Bed 86 in 1922, and planted with varieties of wistaria. Three stone (sandstone) lanterns of varying heights were made by Mr F. J. Read, Outer-Domain Stonemason.[26] Only two of these could be located in 1988, one of them being almost overgrown by a large clump of bamboo.

In Japanese gardens, the smaller lanterns were used to shed light on the paths, the taller ones illuminating the surrounding trees and shrubs.

There are five forms that create a stone lantern. The triangle represents heaven or fire; the circle, water; and the rectangle, the earth. In a religious context, the triangle also symbolises the hands of man or the mirror. Then to these basic forms are added the half-circle or half-moon; an abstraction denoting the wind; and a persimmon-shaped globe for the sky.[27]

In 1983, Sydney's sister-city, Nagoya, presented five fine examples of Japanese stone (white granite) lanterns to the city, on behalf of the efforts of Mr M. Hazumi, Consul-General for Japan in Sydney. Two of these went to Hyde Park and were later vandalised. Three came to the Gardens and were also placed in Bed 88. The Japanese inscriptions say: 'Donated by the city of Nagoya, 1983'.[28]

Sydney Opera House from the Domain, *Jaime Plaza.*

The Arc Glasshouse, Sydney Tropical Centre, *Jaime Plaza.*

Summer view of the city from the Gardens, *Jaime Plaza.*

Frontispiece and music for Garden Palace Opening Ceremony Cantata, *the Trustees, Museum of Applied Arts and Sciences, Sydney.*

Henry Kendall Memorial Seat

Location: South western corner of Lawn 2, Upper Garden

Late art deco style seat, designed by Wilfred Rhodes of the New South Wales Department of Public Works (1939), erected by Saunders (Bondi Quarries Ltd) from working drawings by Charles Wetherburn. The winged horses were carved by L. Bicego. The seat was a gift to the Gardens from Mrs A. Hamilton-Grey in 1940, who admired Kendall's work.

The poet Henry Kendall (1839–82) was born near Milton, New South Wales, and died in Sydney and was buried in Waverley Cemetery. His most famous poem, 'Bell Birds', was very popular with generations of teachers in New South Wales schools:

> The silver-voiced bell-birds, the darlings of day-time,
> They sing in September their songs of the May-time.

The poem is about rainforest:

> By channels of coolness the echoes are calling,
> And down the dim gorges I hear the creek falling.

Kendall had spent part of his youth in Grafton and was obviously sensitive to forests long before others were. After a struggle with alcohol and a breakdown in 1872, he became a guest of the Fagan family near Gosford, which provided inspiration for some of his best-known poems.

In 1879 Kendall wrote the words for the cantata to be sung at the opening of the Sydney International Exhibition (see p.31).

Part I

Songs of morning, with your breath
Sing the darkness now to death —
Radiant river, beaming bay,
Fair as summer shine to-day —
Flying torrent, falling slope,
Wear the face as bright as hope —
Wind and woodland, hill and sea,
Lift your voices — sing for glee!
Greet the guests your fame has won —
Put your brightest garments on.

Part II

Upon the hills that blaze to-day
With splendid dome and spire,
The naked hunter tracked his prey,
And slumbered by his fire.
Within the sound of shipless seas,
The wild rose used to blow
About the feet of royal trees,
A hundred years ago.[29]

To my ear these words have dated and are not representative of Kendall's more enduring work. When sung, however, at an open-air ceremony to mark the centenary of the burning down of the Garden Palace held on 22 September 1982, they sounded wonderful.

The more heartfelt words of the poet are embossed on the curving sandstone back of the seat: 'All my days have been the days of a laborious life, and ever on my struggling soul has burned the fierce heat of this hurried sphere'. A translation of the quotation in Greek (script) at the end of the seat reads: 'I am walking in the air and I am thinking about the sun'.

This is a quotation from Aristophanes' comedy, *The Clouds*, in which Socrates is portrayed as a comic philosopher.

The seat was re-sited and restored in 1967, after the construction of the Cahill Expressway.

Levy Fountain

Location: Near Bed 7, at intersection of pathways near Woolloomooloo Gate, Upper Garden
An art nouveau fountain approximately 4.1 metres square of polished red and white granite, surmounted by a cast bronze statue of a young girl and a heron, with reeds and frogs at the base — by C. B. Birch (1832–93) — a cast of which was exhibited at the Grosvenor Gallery, London, in 1889.[30] The fountain, erected in 1889, was a gift from the Levy family and is an important example of aestheticism/art nouveau in Sydney.

Lewis Wolfe Levy (1815–85) was a businessman and politician who migrated to Sydney in 1840. In 1871 he was elected to the Legislative Assembly for Liverpool and was very active in charitable, educational and religious institutions. He was Director of the Royal Prince Alfred Hospital and the Industrial Blind Institution, and President of the Macquarie Street Synagogue, 1862–74 and 1876–77 (when it closed). He married Julia Solomon in 1845 and had eight sons and seven daughters.

Linda da Chamonix (also see Savoyard Sweep Boy)

Location: Bed 81, Palace Garden, opposite Savoyard Sweep Boy
Linda da Chamonix is a white marble Victorian-era park statue of a young girl with scarf, clogs and musical instrument on a sandstone plinth. Chamonix is in Savoy in the French Alps, near the Swiss and Italian borders.

Linda da Chamonix is the main character in the eponymous opera of Donizetti, written in 1842. She has been described as 'one of those pure, sweet, unsophisticated creatures, who exist only on the stage, and possibly only in opera'.

The musical instrument she is holding is a hurdy-gurdy, which resembles a lute or guitar with strings sounded by a rosined wheel turned with the right hand, and keys to 'stop' the strings played with the left hand.

Following vandalism in the mid-1980s this statue was given a 'silicon bath' treatment and restored to the place where she has been standing for about 100 years. Linda was knocked from her pedestal during the rampage of the evening of 13 July 1989 (when Spring was destroyed) and again taken away for repairs.

Linda da Chamonix, 1989, *Jaime Plaza.*

Lion and Lioness (Sitting Lions)

Location: Adjacent to Bed 38 on either side of Gate (Lion Gate) in the eastern end of the Macquarie Wall
The Lion and Lioness, sitting near the gate at the eastern end of the Macquarie Wall have been a great favourite over the years. Countless children, 'undeterred by the ferocious, or at least intolerant appearance of these beasts ... have sat on the backs. In the course of time the metal surfaces have been polished to a fine smoothness by young posteriors.'[31]

The lions are made from an iron casting on which copper had been deposited by electrolysis. As of 1989 the copper had been worn away, and they were protected from rusting only by a coat of paint. They were removed for copper plating to be returned, resplendent, in 1990. Mrs Nell Felewood wrote to the Gardens in September 1991 to welcome back the lion and lioness. She enclosed a photograph of her granddaughter, Kiri Gerlock, sitting on the lion — a family tradition of an extended family that has been visiting the Gardens annually since 1901.

The Lion and Lioness were presented to the Gardens by the executors of the estate of Edward Sanders in 1912. Sanders was a New South Wales member of parliament and proprietor of a horse-and-waggon carrying business which plied between Homebush and the city. His horse stables were at the city end of Pyrmont bridge.

The lions had been imported from England and stood in the front garden of his Homebush home before they came to the Gardens.[32] They have guarded the gate for such a long period that they have now given their name to it, as well as to Lion Gate Lodge, the Cottage of the Friends of the Gardens.

Love Led Them (Fountain)

Location: Central pond, Pioneer Memorial Garden, Palace Garden
A small bronze fountain figure of a Cupid (young boy with bow and arrows and wings) on a marble plinth in the central pond of the Pioneer Memorial Garden by the sculptor, Paul R. Montford (1868–1938).

Paul Montford was born in London and came to Australia early this century and lived in Melbourne. An important figure in the Australian art scene of the 1930s, his works include the Melbourne Shrine of Remembrance and the statue of Adam Lindsay Gordon in the Spring Street Gardens, Melbourne.

Love Led Them marks the centre of the dome of the Garden Palace (see p.32), previously the precise spot of the foundation stone of a large bronze statue of the young Queen Victoria that had stood in the Palace before the 'melt-down' of 1882, when the great fire destroyed this statue, the Garden Palace, and everything else inside. After the removal of the Queen's Statue foundation (see Rose Garden Pavilion, p.80), the site was developed with soil from the railway paddock at Erskineville and planted with tea-roses in August 1897.[33]

The Memorial Garden to the pioneer men and women, sponsored by the Women's Executive Committee and advisory council of Australia's 150th anniversary celebrations, was officially opened by Her Excellency, Lady Gowrie, on 3 February 1938.

Mare and Foal

Location: Lawn 34, Lower Garden
The bronze Mare and Foal, by the French artist, Arthur J. le Duc (1848–1918) was produced in 1890 (as indicated by the inscription on the base near the mare's hind leg) at Thiebaut Frères' foundry, Paris. Purchased in Paris in 1891, it was donated to the Gardens by the Art Gallery of New South Wales in 1958. This bronze is the only known work by this artist in Australia. The mare's name was La Reyna and the four bronze reliefs around the base of the statue illustrate the history of her training.

Mare and Foal, *c.* 1943, *Mrs Peg Deane.*

On the evening of 5 February 1966 vandals prised the bronze statue of the foal from its base. This was subsequently restored by Mr Alan Ingham, sculptor, for $750.[34]

Unfortunately most of the works of Arthur le Duc, which were gathered in the Torigini Castle Museum (Normandy, France) before World War II, were destroyed by the bombing of Torigini Castle in 1944.[35]

Charles Moore Plaque

Location: A memorial plaque to the memory of Charles Moore was installed on the remaining portion of the old wall that had been incorporated into the new Gardens Restaurant in 1978
The original Refreshment Room was built during the administration of Charles Moore, who was Director of the Gardens from 1848 to 1896. In this extraordinarily long period of 48 years he was responsible for the Victorian-style landscaping and layout of both the Lower Garden and Palace Garden, and responsible for creating the Gardens that we know and love.

Nellie Stewart Memorial

Location: Site of Old Rose Garden
Nellie Stewart was born in Woolloomooloo in 1858. She had a good soprano voice, which she lost in middle life, but was an enduring actress of great popularity who captivated her audiences with her interpretation of both humour and emotion, and brought the spirit of theatre to the screen in *Sweet Nell of Old Drury*. Her autobiography reveals a well-educated and kindly woman of considerable charm. She appeared to be unaffected and free of the jealousies that are often associated with stage life, and held the admiration and affection of Australian playgoers for 50 years.

After she died on 21 June 1932 a Memorial Fund was established by a Mrs W. E. Ardill, which raised enough money for a Garden of Memory. The bronze portrait bas-relief bust of the famous actress was unveiled by Lady Wakehurst on 21 June 1938.

The garden contained some 2,000 roses, including Ma Mie Rosette and Dubarry, the two pinks which were Nellie Stewart's favourite colours. They were officially presented on 20 November 1936 by the Lady Mayoress (Mrs George A. Parkes) to the Minister of Agriculture (Mr Main), who was the Chairman of the Trustees of the Gardens at the time.

The actress's grandson, John Stewart Lynton, presented Mrs Parkes with a basket of oranges at the ceremony. Speaking of his grandmother as 'Ningo', he said he remembered her voice more than anything else. His mother, Mrs Mayne Lynton, was unable to be present as she was in Melbourne, but listened to a radio broadcast of the ceremony. Mrs Parkes then gave a short address in which she spoke of Nellie Stewart's work for charity, including the two matinees in aid of the radium fund at Sydney Hospital, for which she raised hundreds of pounds.

In accepting the garden on behalf of the Trustees of the Botanic Gardens, Mr Main said that it would be one of the most beautiful spots in the Gardens.

After the ceremony, many of the older people present exchanged reminiscences. Miss Eileen Robinson, who was a great friend of the actress, told of the original 'Ma Mie Rosette' frock, of corded silk, which was still in her possession. She recalled also the famous Nellie Stewart bangle, which was made of 20 bangles melted down into one, and worn by her until she died.

The Rose Garden was relocated to the Palace Garden in 1988. A Herb Garden, incorporating a reconstruction of the columned rose arbor around the Nellie Stewart Memorial, is to be established on this site.

Persephone

Location: Anderson Building
Marble bust (1219) 24½ inches by the British sculptor, Marshall Wood, who died in 1882, donated by the Art Gallery of New South Wales in 1958. The daughter of Demeter (the Greek goddess of the harvest), Persephone was abducted by Hades (Pluto) and carried to the nether world, where she was forced to remain for a part of each year — symbolising the seasons and the cycle of nature.

(Governor) Phillip Fountain and Statue

Location: Lawn 11
The Governor Phillip memorial was originally to be located in Centennial Park[36] but after some debate between the Colonial Architect and the Chief Secretary, it was sited in the Gardens.

The massive statue and fountain by Signor Achille

Simonetti, unveiled by the Governor, Viscount Hampden, during the Diamond Jubilee Celebrations of 1897 (i.e., 60 years reign of Queen Victoria), is a wonderful example of high-Victorian exuberance produced for what was then a breathtaking sculptor's fee of £13,000. To put this into some form of perspective, the annual salary for Joseph Maiden, the Director of the Gardens in the year 1896, was £550.

The bronze figure of Governor Phillip, 4.6 metres in height, was cast in Florence. In his left hand he grasps the British flag, while his right holds a half-opened scroll. Within the segments of a circle cut off by bronze reclining figures are marble water basins, with water flowing from the twin nostrils of bronze dolphins. Each marble basin is about 7 metres across and slightly less than a metre deep. The outside walls of the basins are of Bardiglio marble which was chosen because it is impervious to water.

The marble was obtained from Carrara in Italy, where the bronze reclining figures were also cast. The figures represent commerce, agriculture, navigation and mining; the three marble bas-reliefs on the central column represent justice, patriotism and education; the fourth side bears the inscription plate.

At ground level there are four bronze bas-reliefs of Aborigines with boomerang, shield and club, spear and fishing spear respectively.

It is interesting to note that the Cyclops, holding a geological pick (and representing mining in this instance) has not been represented by the artist as single-eyed. Apparently not all classical writers regarded them as such — the Cyclops were generally considered to be the assistants of Hephaestus, the Greek God of Fire (equivalent to the Roman God Vulcan) and as such to have worked on the smelting of metals, thus the connection with mining.

Initially the figures of Mining (Cyclops) and Navigation (Neptune) were to be represented as a contemporary miner in a bush hat and a sailor. The government of the day considered such 'modern' elements to be inappropriate on a classical monument, so Simonetti was directed to alter his half-completed work to the figures of Cyclops and Neptune.[37]

During the construction of the fountain the lawn on the north-eastern side of the statue was 4 or 5 feet above the rest. The surface of the entire lawn was raised to the same level, using soil excavated from the additions to the Treasury buildings in Macquarie Street.

In 1902, J. H. Maiden reported that:

> The basins surrounding this fountain have hitherto presented a somewhat bare appearance. Consequently, some different species of Waterlily (*Nymphea*) have been planted in large unglazed earthenware pans, and also some clumps of Papyrus (*Papyrus antiquorum*, Willd.), in large plant pots, were also deposited in the basins. Already some amount of growth has been made, the former also having produced flowers. The effect is pleasing, softening the general effect of the large marble masses composing the setting of the statue.[38]
>
> The erection of the Governor Phillip statue rendered the small basin fountain at the principal (now Morshead Gate) entrance unnecessary, owing to the larger fountains of the statue — about a hundred yards away — dwarfing the smaller one ... Its removal in 1899 improved the appearance of the entrance, which formerly had a rather cramped appearance.[39]

The Satyr

Location: Bed 129

The Satyr — a life-sized statue of a Pan-like creature, half man and half goat, with a smug expression — was completed by the Australian sculptor, Frank 'Guy' Lynch, in 1924. The artist's brother Joe, a promising young cartoonist, sat for the sculpture and they tied up a goat to help it pose so they could copy its legs.

At its first exhibition in Anthony Hordern's store in 1924 it was praised by art critics, but civic leaders and churchmen expressed shock and demanded that a certain part of the statue be covered with a figleaf. But Lynch, according to his wife Madge, refused because, 'He was a very direct and honest man and did not like to cover anything up — [which] was reflected in his work'.

The statue was completed after World War I and the artist tried to cram every facet of human nature into the face — the heroism, the horror, the happiness and the despair — that he had experienced in the trenches. Its sexuality, however, offended some civic leaders who said it was 'insolent, vicious and animal to the last degree'.

The statue was on exhibition for many years in the 1920s. In 1926 it was bought for the Art Gallery for £250, half the purchase price being subscribed to by the Millions Club, a prominent social group whose members included Dame Nellie Melba. Mr Jack Lang, Premier, was an admirer of the work and wanted a bronze cast to be made so it could go on permanent display, but nothing happened. So it was kept in storage for 50 years and there was an outcry whenever it was brought out.

Finally, permission was given in the 1970s for a bronze cast of the statue to be erected in the Gardens. The Satyr was offered by the Trustees of the Art Gallery of New South Wales as a suitable replacement for Retaliation, which they decided should not have been released. Mrs Lynch paid $3,500 to have the bronze cast made and she said it made her very happy, as it had been her husband's last wish before he died that The Satyr be on permanent exhibition.

The statue was erected in a rocky outcrop in the fern grotto near the Opera House Gate on 1 July 1977.

Joe Lynch had fallen off a ferry near this spot and drowned on the way to a 'bacchanalian' party with a group of artists in 1927. He had been carrying beer bottles in his coat which acted as ballast and his body was never found. The spot had a sentimental value to Guy Lynch, as he always went there after Joe had disappeared. The two brothers had always been very close — their spirits are now fused in bronze and time — and my skin still prickles at the sound of harbour bells.

The more sombre notes of the night Joe drowned were captured by Kenneth Slessor in his poem, 'Five Bells', and in the painting of the same name by John Olsen, commissioned for the opening of the Opera House.

The Sweep Boy, 1989, *Jaime Plaza.*

Savoyard Sweep Boy (also see Linda da Chamonix)

Location: Bed 80, Palace Garden, opposite Linda da Chamonix
White marble Victorian-era park statue of a young sweep boy, standing on a long coil of rope with a 'weight' to plumb chimneys, looking very cold with a straw broom under his arm. Like Linda da Chamonix, the Sweep Boy has been shivering in this spot for more than 100 years, until the evening of 12 December 1990, when he was knocked off his pedestal.

The Savoyard Sweep Boy was one of the characters of the opera *Linda da Chamonix*, written by Donizetti in 1842. The Savoyards are the natives of Savoy.

The Seasons

From early photographs and etchings we know that four marble statues of the Seasons and two marble urns graced the terrace steps on the main pathway (the stone coming from Pyrmont quarries) leading from the Middle Garden to the Palace Garden in the early 1880s. These statues were imported from Italy in 1879[40] by Mr Job Hanson who won the First Degree of Merit award for his exhibit of Statues and Monuments (tombstones — not to be confused with obelisk (q.v.)) associated with Garden Palace — external display No. 18 in the International Exhibition of 1879.

The exhibition consisted of 'four statues Spring, Summer, Autumn, and Winter, and two monuments (gravestones), one of marble, the other of red granite'.[41] The statues were reported to be 'of pleasing design and well executed'.[42]

A photograph of the period has now established beyond doubt that the statues were displayed in situ on the stairs, with the two monuments adjacent (on the higher ground).

Following the fire which destroyed the Garden Palace and 'the colonial collection of statuary'[43] on 22 September 1882, the statue Spring lost an arm and 'one of the massive marble flower vases standing on the eastern stairs came to grief owing to clumsy handling in removal'.[44] In his *Guide to the Botanic Gardens*, Maiden reports that: 'The left-hand top figure of Spring came to grief during the fire, through contact with the falling debris from the burning building; thus it shows marks of repair'.[45]

Two Seasons on Palace Garden Steps, 1895, *RBG Library*.

The fact that Spring was damaged during the fire probably indicated that the statue was at the top of the steps and closer to the east-west wing of the building. There were later reports of Autumn having been damaged as well. A photograph taken just after the fire shows that both statues at the top of the stairs had been removed.

The other statuary and marble ornaments around the grounds were not damaged by the fire.[46]

The eight marbles reported as being imported from Italy in 1883 also included the four seasons.[47] These marbles, almost certainly from the studio of Charles Francis Summers[48] may be readily identified by the newspaper description:

> Spring, a fair maiden with her lap full of flowers; Summer, with her right hand encircling a sheaf of wheat and cornflowers, while her left hand grasps a sickle; Autumn, a joyous Bachanal, with one hand holding a bunch of grapes while the other tosses aloft a wine-beaker; and Winter, a wrinkled elder clasping a frost-killed bird to his bosom.

An additional statue of Spring came to the Gardens in 1957 from the estate of James Stedman (see Chapter 7).

At some stage the six seasons (the four seasons plus at least two of the earlier four) were dispersed throughout the Gardens. More recently Spring went to Bed 84, Summer to Bed 56 near the Restaurant (both now in 'graveyard' behind Succulent Garden), Autumn to Bed 97 and Winter to Lawn 16 near the steps (now also in 'graveyard').

The only one currently in the Gardens is Autumn (on the Palace Garden Steps).

Autumn

Location: Palace Gardens Steps, Main Pathway
This is the restored statue of Autumn that previously resided in Bed 97. In the Annual Report of 1905 J. H. Maiden records that the statue Autumn required repairs.[49] He also noted that this same statue had required repairs 20 years previously, when it was damaged by falling debris from the eastern tower of the Garden Palace during the fire of 1882. This information is disputed by the article in the *Illustrated Sydney News*, which said the statue Spring was damaged. Maiden was not Director at the time of the fire and would have been relying on memory or notes in brown paper bags[50] — or reporting a second-hand story.

Given that experience has taught me that newspapers are not always totally accurate, and given the evidence of the photograph showing both statues at the top of the stairs having been removed after the fire, leads me to suspect that *both* Spring and Autumn may have been damaged.

I have inspected the existing statue of Autumn and there is no evidence of damage, which is more evidence that this statue is one of the Summers group that arrived in 1883, after the fire.

Solander Bust

Location: Bed 13

In 1982, to mark the 200th anniversary of Solander's death, the Royal Botanic Gardens constructed a special garden in his honour featuring examples of the species collected during his visit to Australia in 1770.

Daniel Carl Solander (1733–82) was born, raised and educated in a Lutheran manse in Pitea, Sweden. Later, at the University of Uppsala, he was a star pupil of the great naturalist, Carl Linnaeus. In the late 1750s he came to England, where from 1763 to 1768 he organised the natural history collection of the British Museum using the Linnaean system of classification, and became a lifelong friend of Joseph Banks.

At the time he joined the Cook voyage (1768–71) as part of the Banks 'suite', he was one of the better naturalists, if not the best naturalist, in England.

There was another who is reputed to have travelled with Banks and Solander on the *Endeavour*, a ship's boy, one Peter Hibbs, an ancestor of mine who later returned to Botany Bay on the *Sirius*.[51] I cannot now read the journals of Cook's voyage without a vicarious pang — that he was there and witnessed those historic events with his own eyes — as if I were a ghost on the ship. Peter Hibbs sailed with Flinders in 1798 as Colonial Master of the supply ship *Norfolk* on the voyage that proved Van Dieman's Land (Tasmania) was indeed an island.

When Banks and Solander returned to England they were feted by London society. Consequently the published results of the voyage were quite small. However, the years between 1773 and 1782 were busy ones for Solander. He continued his work at the British Museum; helped Banks with his projects; was employed by the Duchess of Portland; and aided John Ellis, Doctor Fothergill, and others in their scientific projects. In addition to these occupations, Solander was actively involved in the affairs of the Royal Society, the Royal Society Club, at least one coffee house club and a Society of Physicians.

A massive publishing venture, consisting of copper plates of illustrations of plants collected on the voyage was commenced, but destined to wait some 200 years before completion (*The Banks Florilegium*).

Daniel Solander died in London in May 1782.

The pewter bust of Solander, by the Swedish sculptress, Tora Ceder, who also came from Pitea was commissioned by the Swedish Embassy in Canberra. The statue, mounted on a small granite rock, and the garden were officially dedicated by His Majesty King Carl XVI Gustaf of Sweden on 31 March 1982.

Sundial

Location: Previously in Lower Gardens and surrounded by an iron fence, it disappeared some time ago, to be repurchased on 8 April 1990.

This sundial, presented to the Gardens in 1857, is believed to be the first horizontal sundial shown in Sydney.

Mrs Valmai Pearse, corresponding with Dr Penny Farrant, said their family legend was that the sundial was engraved by her grandfather, Mr Thomas Salisbury, as recorded in his obituary: 'There are many evidences of his work in and around the city, including the sundial in the Botanic Gardens . . . the brasses of St. Andrew's Cathedral, and the heraldic devices at the Sydney University'.[52] Young Thomas is reported to have told his mother 'how disappointed he was not to be allowed to place his name on his pride and joy — the sundial for the Botanic Gardens — but that he had managed to engrave a small T.S. on the rim!'

The inscription in the sundial reads:

This sundial
was presented to
The Botanical Gardens Sydney
by the maker
Archibald Bolan
AD 1857

It is assumed that Thomas Salisbury worked for Archibald Bolan. The incription on the rim could not be located in 1990.

Thomas Salisbury was also reputed to have done an engraving on the shell of the tortoise in the Gardens, but an inspection of the remaining preserved tortoise shell in 1990 could not locate any inscription. There were,

The Sundial, Lower Garden, 1887, *RBG Library*.

however, at least three tortoises in the Gardens (see p. 31) so he could have engraved one of the others.

As was the case with the Wishing Tree, others were to claim a link with the sundial in the Botanic Gardens. An obituary for a William Wilson (no relative) who died in 1922 aged 92 years, said he 'was responsible for engraving the sundial at present to be seen in the Botanical Gardens'.[53] This must have been the same sundial, as the 'Grantham' sundial (see p.133) was not presented to the Gardens until 1927.

The sundial was octagonal in shape, made from black Italian marble, and mounted on a sandstone pedestal. Later to be protected by an iron fence, it stood at the intersection of the pathways surrounded by Beds 93, 94, 95 and 96.

At some unidentified period the sundial went missing until it was 're-discovered' in an antique shop at Windsor, and finally purchased for $5,200 (with the assistance of a benefactor) at an auction sale on 8 April 1990.

At the time of writing it was the Director's intention to have a cast of the sundial made for external use, so the original could be better preserved. The current thinking was that the replica will be placed somewhere in the Herb Garden to be developed on the site of the Old Rose Garden. The original is currently on display at the switchboard.

Terracotta Vases

Location: Base of steps to Visitor Centre
Two terra cotta vases were selected in 1873 by Dr George Bennett at the establishment of Doulton and Watts, Lambeth, London. A sum of £100 was voted for their purchase in 1878. They were obtained in 1879 and placed at the base of the steps leading into the Museum and Lecture Hall (now Visitor Centre and Gardens Shop).

As of 1988 only one remains. Apparently it was cracked and had been painted the colour of the building; the other, unpainted, is in storage in the Outer Domain Fuel Shed.

Dr George Bennett had trained as a surgeon in London and settled in Sydney, in part influenced by the beauty of the spring flowers, to become an influential naturalist.[54] He was a Director of The Australian Museum and associated with these Gardens, and a founding member of the Acclimatisation Society (see p. 30).

The painter Marianne North met Bennett when he was about 80 years old and he escorted her around the Gardens, showing her 'with great delight how much bigger the *Ficus Bennettii* [*sic*] was than the *Ficus Moorii* [*sic*]'.[55]

Venus Italica

Location: A superb 5-foot marble replica of the Venus Italica by the Italian sculptor Antonio Canova (1757–1822) in Bed 86
Venus, the Roman goddess of love, is regarded as having sprung from the foam of the sea. Here she is seen returning from her bath.

This replica was purchased in Melbourne in 1890 and transferred to the Gardens from the Art Gallery of New South Wales in 1958.[56] She was located under a fine specimen of *Ficus maclellandii* var. *rhododendrifolia* in the Lower Garden, in a bed of *Clivias* (Bed 86). On the evening of 16 January 1989, she was decapitated.

A member of the Friends of the Gardens, Mrs Rosemary Wells, discovered the head in the garden near the Access Road on 23 February. It is believed that, following the publicity associated with the theft, the head may have become too hot to handle. Venus was then sent to T. A. Taylor and Son for cleaning, waterproofing, and re-affixing of the head.

Venus Fountain in Lower Garden, 1887. Note clam shell at base of statue, *RBG Library*.

'Venus' Fountain

Location: Main Pond, Lower Garden

This zinc alloy (with lead intrusions) cast of a Venus Fountain with cupids and dolphins at her feet may be found on a rock base in the large pond of the Lower Garden. There are double fountain sprays from each of the two dolphins, and shell fountains from each of the two cupids.

A dated photograph in the Gardens' archives indicates this statue was erected prior to 1880, probably in 1879 leading up to the International Exhibition. At this stage the base of the fountain was surrounded by weathered rocks and large clam shells. These were later removed and replaced by a sculptured crocodile (alligator?). The base of the fountain is currently unadorned.

The badly deteriorating statue was removed 8 August 1990, to be cast in bronze by Fineart Bronze Foundry Pty Ltd at Brookvale.

Chronological Checklist of Existing Statues, Fountains and Memorials in the Gardens

1796	Joseph Gerrald 'Scottish Martyr' buried in Middle Garden
1812	Macquarie Wall
1844	Allan Cunningham Obelisk
1857	Sundial
c. 1879	The Boxers
c. 1879	Huntsman and Dogs Statue
c. 1879	Linda da Chamonix
c. 1879	Savoyard Sweep Boy
1879	Terra Cotta Vase
c. 1879	Venus Fountain
c. 1879	Bird Bath and Fountain
1883	Boy Extracting Thorn
1883	Autumn
1889	Levy Fountain
1897	Governor Phillip Fountain and Statue
1912	Lion and Lioness (Sitting Lions — from estate Edward Sanders)
1912	Japanese Lanterns (Sandstone)
1921	Venus Italica
1938	Love Led Them
1938	Nellie Stewart Memorial
1940	Henry Kendall Memorial Seat
1943	Choragic Monument of Lysicrates (Erected 1870 Clarens — from estate Sir James Martin)
1946	I Wish
1952	Darnell-Smith Memorial Seat
1952	First Farm Memorial
1957	Spring
1958	Cometh Up as a Flower (Produced 1878)
c. 1958	Persephone
1959	Mare and Foal
1977	The Satyr (Original produced 1924)
1978	Charles Moore Plaque, Gardens Restaurant
c. 1980	Allan Cunningham's tombstone (1839) incorporated into new Herbarium wall (Robert Brown Building)
1981	Bird Bath (Sydney Garden Club)
1982	Solander Bust
1983	Japanese Lanterns (white granite)
1988	Banks Bust (Robert Brown Building)
1988	Dolphin
1988	Phillip Wall (see Chapter 6)

6

The Plaque near the Ladies Toilet

Alphabetical Checklist of Statues, Fountains and Memorials in the Domain

Almost Once (Redhead/Blackhead)

Location: Eastern Lawn between Art Gallery of New South Wales and Sir John Young Crescent
A recent Brett Whiteley statue consisting of a pair of large 'Redhead' matchsticks — one live and one burnt — on a concrete base. The statue is made of blackbutt timber and fibreglass. It was conceived in 1968 and finished in 1991, and has had many name changes.

Bronze Horses (miniature pair)

Location: On plinths on either side of steps opposite Art Gallery of New South Wales
These two fine miniature bronze equestrian statues opposite the Art Gallery of New South Wales were donated by the Art Gallery Trustees,[1] and erected in November 1964.[2] Completed in 1702 and 1709 by the sculptor Antonius Coyzevox, they both carry the transcription 'LUGO.SCUL.REG.FECIT' (of Lyon (French city of Lyons) the King's Sculptor, made it).

Antonius Coyzevox (1640–1720) the French sculptor of Spanish extraction, was born in Lyons. He went to Paris in 1657 and was appointed Sculptor to the King in the same year. After a later period in Germany he returned to Paris and produced many statues for the royal gardens at Marli and Versailles. A bronze equestrian statue of Louis XVI at Rennes was destroyed during the revolution.[3]

Robert Burns

Location: Near Domain Lodge, Art Gallery Road
In the Annual Report of 1904 it is recorded that a handsome and massive granite pedestal had been erected near the corner of the Rosarium, soon to support a bronze statue of the Scottish poet, Robert Burns.

The pedestal and base steps was 14 feet high. The three base steps (the lowermost being 18 feet square) were made of Bowral trachyte, partly polished; the pedestal of Melbourne granite, mostly in the rough, with one polished course in the centre, cost £600. The site, close to the carriage road (Art Gallery Road) was chosen so the statue faces the afternoon sun, and was used for many years for annual Burns Day Celebrations.

The statue, a bronze by Frederick Pomeroy, 'one of the most pleasing in the State' — some nine feet six inches (2.9m) high, giving a total height of twenty-three feet six inches (7.2m) — was commissioned by the Burns Memorial Committee, and cost £1000. It was unveiled on 30 January 1905 (Anniversary Day) by the State Governor, Sir Harry Rawson; assisted by the Federal Prime Minister; the Premier of Queensland; the Lord Mayor of Sydney; the Burns Memorial Committee; and other notables, in the presence of a very large gathering,

Miniature bronze horse by Antonius Coyzevox, 1702–09, 1989, *Jaime Plaza.*

in the rain — 'the day and the hour of unveiling were amongst the wettest on record'.

There's a delightful poem by Henry Lawson, 'Bobbie's Statue', which was composed in February 1905 and published in *The Bulletin*, 23 February 1905. From a combination of the timing and invective I assume Lawson was part of the larger gathering in the rain, who hadn't been 'officially' invited. His spleen, and a certain amount of professional jealousy are directed at the 'toffs' and their adoration for the dead poet — when they wouldn't have invited the live, shabby and penniless poet (Lawson included) into their houses had he appeared before them. It would have been very easy for Lawson to feel sorry for himself, although he managed a statue in the Domain for himself after his own death (p.112).

Bobbie's Statue

Grown tired of mourning for my sins —
  And brooding over merits —
The other night with aching heart
  I went amongst the spirits;
And I met one that I knew well:
  'O Scotty's Ghost! is that you?
And did you see the fearsome crowd
  At Bobbie Burn's statue?

'They hurried up in hansom cabs,
  Tall-hatted and frock-coated;
They trained it in from all the towns,
  The weird and hairy-throated;
They spoke in some outlandish tongue,
  They cut some comic capers,
And ilka man was wild to get
  His name in all the papers.

'They showed no sign of intellect,
  Those frauds who rushed before us;
They knew one verse of *Auld Lang Syne* —
  The first one and the chorus.
They clacked the clack o' Scotlan's Bard,
  They glibly talked of ''Rabby'';
But what if he had come to them
  Without a groat and shabby?

'They drank and wept for Rabbie's sake,
  They stood and brayed like asses
(The living bard's a drunken rake —
  The dead one loved the lasses);
If Robbie Burns were here, they'd sit
  As still as any mouse is;
If Bobbie Burns should come their way,
  They'd turn him out their houses.

Robert Burns, 1989, *Jaime Plaza.*

'O weep for bonny Scotland's Bard!
And praise the Scottish nation,
Who made him spy and let him die
Heart-broken in privation:
Exciseman, so that he might live
Through northern winters' rigours —
Just as in southern lands they give
The hard-up rhymer figures.

'We need some songs of stinging fun
To wake the States and light 'em;
I wish a man like Robert Burns
Were here to-day to write 'em!
But still the mockery shall survive
Till Day o'Judgement crashes —
The men we scorn when we're alive
With praise insult our ashes.'

And Scotty's Ghost said, 'Never mind
The fleas that you inherit;
The living bard can flick 'em off —
They cannot hurt his spirit.
The crawlers round the poet's name
Shall crawl through all the ages;
His work's the living thing, and they
Are fly-dirt on the pages.'

Governor Bourke (not on Trust Land)

Location: Once near current site of Shakespeare Group, now near Mitchell Library

On a brilliant sunny day on Monday, 11 April 1842, the largest crowd ever assembled in Sydney (estimated between 7,000 and 10,000) attended the unveiling of a life-sized bronze statue of Lieutenant-General Sir Richard Bourke, Governor of New South Wales from 1831 to 1837, in the presence of Sir George Gipps, the current Governor.[4] The occasion was celebrated by a general holiday in the town and Governor Gipps declared in his speech that 'the view presented from this spot equalled in loveliness any scene in the known world'.[5]

The statue, the first of its kind to be commissioned in Australia, was paid for by a subscription from a committee of friends and private donations. Sir Roger Therry was the Honorary Secretary of the committee responsible for erecting the statue. Its cost was £2,000, and the estimated cost of the pedestal, etc., was £400. At the time the names of the subscribers were published in *The Australian* newspaper, the total was £1,809 1*s*. 4*d*., from over 800 people in sums from £50 downwards. By June 1839, the amount subscribed was £2,191 9*s*. 9*d*., and the balance (£208 10*s*. 3*d*.) had been promised.[6]

Mrs Charles Meredith reported in 1839 that:

> A group of graceful weeping-willows overhang a pretty shady pool, where a statue, by an English sculptor [Westmacott, I think — N.B. Attributed to Bailey in *S.M.H.* 20 June 1846], is now erected to General Bourke, formerly governor of New South Wales. It had not arrived when we left Sydney, nor I should have much rejoiced to see the first specimen of high art which the colony has obtained, placed in so lovely and, with us, so favourite a spot.[7]

Designed by E. H. Bailey,[8] the statue was to have been brought out free of charge on one of the convict ships but problems arose due to the discontinuation of transportation.[9] The Colonial Secretary then asked if the statue be transported free of charge on one of the ships doing government business.[10] It is not known who subsequently paid the freight, if indeed any was charged. The statue obviously arrived.

Originally it was to have been erected in the centre of Hyde Park, 'where the two main roads cross each other'.[11] Since at one stage it was intended that Macquarie Street be extended south through the present Hyde Park, the proposed spot 'where the two main roads cross each other' would have been the intersection of this intended extension of Macquarie Street with the present Park Street.

This proposal was abandoned and the statue was erected close to its present site at the old Bent Street entrance to the Domain in 1842,[12] roughly corresponding with the spot where the Shakespeare Group now stands, and flanked by two 'Russian guns'.[13]

The crowd at the unveiling was reportedly very well behaved and there was not the 'slightest impropriety'. The occasion had the air of a 'general holiday' and a 'forget and forgive' attitude prevailed, even among those who had been prominent in opposing Bourke's policies.[14]

In his estimate for the Government Domains in 1858, Charles Moore, Director of the Gardens, referred to a 'road extending from the statue of Sir R. Bourke, to the Domain entrance from Hyde Park, eighty-seven yards'. The statue stood for a long time in this spot, surrounded by a picket fence and was later flanked by two cannon, near an original stand of *Casuarina glauca* (swamp she-oak).[15]

These cannon were captured by British troops after the siege of Sebastopol in 1854/5. They had been mounted on the ramparts guarding the dockyards of the city of Sebastopol, the Russian naval base in the Black Sea. When the city fell the guns were taken back to England and France, and pairs of guns were sent to the cities of the British Empire.[16]

The Crimean War, resulting from a 'quarrel between Russia and [the] Catholic Powers',[17] generated great carnage through bureaucratic incompetence and bad military decisions, with traditional forms of attack against an entrenched and superior technology. The Light Brigade charged similar guns at Balaclava, as captured in Alfred (Lord) Tennyson's 'The Charge of the Light Brigade'.[18] In all, approximately 100,000 English lives were lost. The French lost an equivalent number of men; the Russians lost even more.[19]

The guns were later moved to Centennial Park (see Chapter 8, p.137), where they may still be seen.

The inscription on the statue, which now stands outside the Mitchell Library, reads:

THIS STATUE
OF
SIR RICHARD BOURKE, K.C.B.
IS ERECTED BY THE PEOPLE OF NEW SOUTH WALES
TO RECORD HIS ABLE, HONEST, AND BENEVOLENT ADMINISTRATION
FROM 1831 TO 1837. SELECTED FOR THE GOVERNMENT AT A PERIOD OF SINGULAR DIFFICULTY
HIS JUDGEMENT, URBANITY, AND FIRMNESS JUSTIFIED THE CHOICE.
COMPREHENDING AT ONCE THE VAST RESOURCES
PECULIAR TO THIS COLONY,
HE APPLIED THEM, FOR THE FIRST TIME, SYSTEMATICALLY TO ITS BENEFIT.
HE VOLUNTARILY DIVISTED HIMSELF OF THE PRODIGIOUS INFLUENCE
ARISING FROM THE ASSIGNMENT OF PENAL LABOUR, AND ENACTED
JUST AND SALUTARY LAWS FOR THE AMELIORATION OF PENAL DISCIPLINE.
HE WAS THE FIRST GOVERNOR WHO PUBLISHED SATISFACTORY ACCOUNTS
OF THE PUBLIC RECEIPTS AND EXPENDITURE.
WITHOUT OPPRESSION, OR DETRIMENT TO ANY INTEREST,
HE RAISED THE REVENUE TO A VAST AMOUNT, AND FROM ITS SURPLUS,
REALIZED EXTENSIVE PLANS OF IMMIGRATION.
HE ESTABLISHED RELIGIOUS EQUALITY ON A JUST AND FIRM BASIS,
AND SOUGHT TO PROVIDE FOR ALL, WITHOUT DISTINCTION OF SECT,
A SOUND AND ADEQUATE SYSTEM OF NATIONAL EDUCATION.
HE CONSTRUCTED VARIOUS PUBLIC WORKS OF PERMANENT UTILITY.
HE FOUNDED THE FLOURISHING SETTLEMENT OF PORT PHILLIP
AND THREW OPEN THE UNLIMITED WILDS OF AUSTRALIA
TO PASTORAL ENTERPRIZE.
HE ESTABLISHED SAVINGS BANKS, AND WAS THE PATRON OF
THE FIRST MECHANICS' INSTITUTE. HE CREATED AN EQUITABLE TRIBUNAL
FOR DETERMINING UPON CLAIMS TO GRANTS OF LANDS.
HE WAS THE WARM FRIEND OF THE LIBERTY OF THE PRESS, HE EXTENDED
TRIAL BY JURY AFTER ITS ALMOST TOTAL SUSPENSION FOR MANY YEARS.
BY THESE AND NUMEROUS OTHER MEASURES
FOR THE MORAL, RELIGIOUS, AND GENERAL IMPROVEMENT OF ALL CLASSES,
HE RAISED THE COLONY TO UNEXAMPLED PROSPERITY;
AND RETIRED AMID THE REVERENT AND AFFECTIONAL REGRET
OF THE PEOPLE, HAVING WON THEIR CONFIDENCE BY HIS INTEGRITY,
THEIR GRATITUDE BY HIS SERVICES, THEIR ADMIRATION BY HIS PUBLIC TALENTS
AND THEIR ESTEEM BY HIS PRIVATE WORTH.

Desert Corps Horse Memorial

Location: Macquarie Street, on fence near Morshead Gate

Erected on 25 April 1950 (Anzac Day) by friends and members of the Desert Mounted Corps on the fence opposite the State Library of New South Wales in memory of the horses which carried them over the Sinai Desert into Palestine 1915–18.

The importance of the site opposite the State Library was that 'the majority of members were country men, and a less prominent location would not be so valued nor would the Memorial be so conveniently seen' and for many years the men of the Desert Corps fell in for their Anzac Day parade at this spot.

The panel, framed by a series of sandstone blocks, comprises two metal plaques with a bronze bas-relief by J. B. Dunlop of three horses with ammunition pouches around their necks, and reins held by a trooper in short sleeves, hat and boots, with ammunition pouches across his chest.[20] The inscription reads: 'They suffered wounds, thirst, hunger almost beyond endurance but never failed. They did not come home. We will never forget them.'

The Memorial was moved to its present location in the mid-1960s, after the construction of the Cahill Expressway.

(Aquarius) Frazer Drinking Fountain (not on Trust Land)

Location: Traffic Island at southern end of Art Gallery Road

A monumental Victorian sandstone drinking fountain, 3 metres square × 5 metres high (approximately), with griffins and pink granite basins inscribed on opposite sides with the aquatic references, Aquarius and Arethusa — the latter referred to by Shelley in his poem,

> Arethusa arose
> From her couch of snows
> In the Acroceraunian mountains —

This is one of the Frazer Fountains, the other one now being in Hyde Park south, opposite Sydney Grammar School.

The York Street merchant, John Frazer, arrived in Sydney in 1840 from Dunmore, Ulster, as a boy of 14, and prospered. The Frazer Fountains, at the entrances to Hyde Park and the Domain, were his gifts to the city.[21]

The inscription records that it was presented by 'John Hardie Mayor 1884 to his fellow citizens', produced by Thomas Sapsford, city architect, and L. Beveridge, sculptor. It is in poor repair and tarnished by city grit, and the drinking fountains are no longer working.

King Edward VII (not on Trust Land)

Location: Traffic Island west of Conservatorium of Music

This bronze 8.5 metre high equestrian statue of King Edward VII by Sir Thomas Brock was unveiled by the Lieutenant Governor, the Hon. Sir William Portus Cullen, on Empire Day, 24 May 1923, to commemorate his reign, 1901–10.

'The over life size figure of the king in plumed hat and uniform, seated on a prancing horse' is mounted on a sandstone plinth. Relief bronzes at each end of the plinth depict allegorical female figures, one pair flanking a shield emblazoned with an emu signifying 'Peace and Progress', the other pair flanking a shield emblazoned with two hands clasping, signifying Unity and Empire.[22]

Just prior to the unveiling the *Sydney Mail* described the scene as follows:

> This beautiful scene — just off Macquarie Street — is near the Conservatorium of Music and only a few minutes' walk from the heart of Sydney. The statue is Sir Bertram Makenal's [Renée Free attributes the work to Sir Thomas Brock[23]] magnificent representation of King Edward VII, father of King George, whose birthday has just been celebrated. To the left is the stately residence of the Governor of New South Wales standing in spacious grounds. Through the trees may be obtained charming glimpses of the harbour. The view, if extended to the right, would include the lovely Botanic Gardens, of which Sydney people are proud.[24]

Fleet Steps and Queen Elizabeth II Memorial

Location: Eastern side Farm Cove, Mrs Macquaries Point

A permanent flight of 108 steps was erected in 1908 by the Government Architect, just prior to the visit of the Great White American Fleet, where the American Admiral, officers and bluejackets first set foot on Australian soil. The decorations for the occasion were elaborate, including flag-poles, bunting, greenery and hot-house plants.

Fleet Steps, 1989, *Jaime Plaza.*

On 9 September the new Governor-General, the Earl of Dudley, made the same spot his official landing place when a procession of carriages moved from the head of the stairs to the Domain cricketground (Phillip Precinct), where the American Reception Pavilion still stood.

The swearing-in ceremony was performed at the foot of the stairs on the Federation Stone, a handsome octagonal piece of polished Australian granite, which was brought from Centennial Park for the purpose. It was at his same spot that the first Governor-General, Lord Hopetoun, had landed at Farm Cove on 1 January 1901. Five months later the Duke and Duchess of York landed there.[25]

Appropriately in 1954, this was the place chosen where Queen Elizabeth II first set foot on Australian soil (asphalt), as recorded in the adjacent monument whose inscription reads: 'At this spot Queen Elizabeth II first landed on Australian soil on 3rd February, 1954'.

In addition to the Fleet Steps some additional 242 steps, consisting of short flights from two to a dozen or more steps in each, have been worked and laid, or hewn out of solid rock, up the steep slopes near Mrs Macquaries Chair by the Domain mason, with the idea of fixing seats at the various good lookout positions.[26]

Matthew Flinders (not on Trust Land)

Location: New State Library of New South Wales facing Macquarie Street

A 5.5 metre bronze figure by English sculptor William Robert Cotton of Matthew Flinders in his eighteenth-century naval uniform of knee breeches and tail coat, with a sextant in his left hand, mounted on a stepped and shouldered granite plinth.[27] The inscription reads:

> Matthew Flinders
> 1774–1814
> Captain in the Royal Navy
> He explored and skilfully charted a large portion of the Australian coast during the years 1796 to 1803.

The sculpture was unveiled on 22 October 1925 by the Governor, Sir Dudley de Chair.

> Some years before, Ifould, the principal Librarian, had approached Flinders' grandson, Professor Flinders Petrie, a noted Egyptologist, asking him to consider bequeathing his collection of Flinders papers to the Mitchell Library. Petrie agreed to donate the collection, on condition that a substantial monument to his grandfather was erected.[28]

The Flinders circumnavigation of Australia is central to the history of Australian botany, and hence to the Gardens, because it carried the great botanist Robert Brown and the artist Ferdinand Bauer. Brown later named the genus *Flindersia* to honour his captain and friend. Both *Flindersia australis* and *Flindersia schottiana* are growing in the Gardens.

On the ground beside the statue is a brass plaque with incised letters and the line drawing of a cat, erected on 11 November 1985 by the North Shore Historical Society with a quotation from Flinders (1809), in memory of his cat Trim:

> Trim, the best and most illustrious of his race and most affectionate of friends, faithful of servants and best of creatures. He made the tour of the globe and a voyage to Australia which he circumnavigated, and was ever the delight and pleasure of his fellow voyagers.

Henry Lawson

Location: Woolloomooloo Precinct of the Domain

After Lawson's heartfelt reaction to the unveiling of the statue to Robert Burns (see p.107), it is ironic that he was to have a statue erected to his own memory, nine years after he died.

Henry Lawson, probably Australia's most famous poet and short story writer, died a pauper in 1922 and was given a State Funeral. He was buried at Waverley Cemetery, where Henry Kendall is also interred. 'A week

The Sweep Boy, *Jaime Plaza.*

Lion and Lioness, *Jaime Plaza.*

Boy Extracting Thorn, *Jaime Plaza.*

I Wish (statue), *Jaime Plaza.*

earlier the cuff-and-collar push crowding the cathedral for his funeral would have crossed the street to avoid being bitten by [him]'.[29]

Funds for the statue were provided by people from all walks of life following a public meeting convened in 1922 by Alderman W. P. McElhone, Lord Mayor of Sydney. Initially only £300 of the £1,500 required for the statue was raised in 12 months.

Mr W. Ifould, State Librarian, was chairman of a new committee responsible for erecting the statue. The bronze, featuring Lawson standing next to a swagman and his dog — two favourite characters of many of his stories — was sculpted by George (G. W.) Lambert, ARA — an old friend of Henry Lawson, son of an American engineer, born in what was then St Petersburg (and is once more), and brought to Australia as a boy. He died shortly after the sculpture had been completed and cast by A. B. Burton of London in 1930.

The statue, on a simple square trachyte pedestal — the lower plinth being smooth, the upper rock faced — was unveiled by the Governor of New South Wales, Sir Philip Game, on 28 July 1931, in the middle of the Great Depression, when the homeless lived in the 'caves and cliff shelters of the Domain' and were hounded by the police.[30]

Henry Lawson, 1989, *Jaime Plaza.*

Members of Lawson's family were invited to the ceremony. In the context of what was to come, it is interesting to note that Henry Lawson's brother-in-law, Jack Lang, was not present. The 'Lang Plan' of February 1931, for withholding interest payments on British loans, and reducing interest on domestic ones, had split the Labor Party and provoked the formation of the 'New Guard'. Lang was dismissed by Sir Philip Game on 13 May 1932.

At the unveiling ceremony Mr Ifould is reported as saying: 'No memorial fund started in Australia has had more subscribers . . . many thousands of school children, shearers, shed workers and members of the Australian Workers' Union contributed. Small coins were sent in from every school in the State.'[31]

This romantic memorial with its poet, swaggie, post-and-rail, and dog now shaded by a Cabbage gum, *Eucalyptus amphifolia*, near where Henry Lawson and his wife were often seen to stroll,[32] always reminds me of my grandfather and his era — not that he was a swaggie although he actually wore corks around his hat to keep the flies off his face!

Life for Henry Lawson, however, was far from romantic. On several occasions he was locked up in Darlinghurst gaol — once as a 'voluntary inebriate', and another time for being unable to pay his alimony — and the kids at North Sydney used to throw stones at him when he came home drunk at night.[33]

Edmund Campion made the point[34] that the statue holds out its hand, as if for a beer, and that from time to time someone manages to climb up and give him a can of Foster's to hold. Lawson may have slept rough in the Domain at one stage of his life when he was down and out. If not, he certainly would have observed those who did in the Depression of the 1890s.

For Heritage Week 1984 we organised for John Derum to read Lawson's verse under his statue, and Chris Shaw sang some of his poems set to music. Ploughing through the *Collected Works*, in a poem called 'Says You', I came across an intriguing couplet 'rain/Domain' which tended to support the theory that he was writing from experience.

Says You

When the heavy sand is yielding backward from
your blistered feet,
And across the distant timber you can see the
flowing heat;

When your head is hot and aching, and the shadeless plain is wide,
And it's fifteen miles to water in the scrub the other side —
Don't give up, don't be downhearted, to a man's strong heart be true!
Take the air in through your nostrils, set your lips and see it through —
For it can't go on for ever, and 'I'll have my day!' says you.

When you're camping in the mulga, and the rain is falling slow,
While you nurse your rheumatism 'neath a strip of calico,
Short of tucker or tobacco, short of sugar or of tea,
And the scrubs are dark and dismal, and the plains are like a sea;
Don't give up and be downhearted — to the soul of man be true!
Grin, if you've a mate to grin for! grin and joke, and don't look blue;
For it can't go on for ever, and 'I'll rise again!' says you.

When you've tramped the Sydney pavements till you've counted all the flags,
And your flapping boot-soles trip you and your clothes are mostly rags,
When you're called a city loafer, shunned, abused, moved on, despised —
Fifty hungry beggars after every jobs that's advertised —
Don't be beaten! Hold your head up! To your wretched self be true;
Set your pride to fight your hunger! Be a man in all you do!
For it can't go on for ever, and 'I'll rise again!' says you.

When you're dossing out in winter, in the darkness and the rain,
Crouching, cramped and cold and hungry, 'neath a seat in The Domain,
And a cloaked policeman stirs you with that mighty foot of his —
'Phwat d'ye mane? Phwat's this? Who are ye? Come, move on — git out av this!'
Don't get mad; that's only foolish; there is nothing you can do,
Save to mark his beat and time him — find another hole or two;
But it can't go on for ever — 'I'll have money yet!' says you.

Don't you fret about the morrow, for sufficient to the day
Is the evil (rather more so). Put your trust in God and pray!
Study well the ant, thou sluggard. Blessed are the meek and low.
Ponder calmly on the lilies — how they idle, how they grow.
A man's a man! Obey your masters! Do not blame the proud and fat,
For the poor are always with them, and they cannot alter that.
Lay your treasure up in heaven — cling to life and see it through!
For it cannot last for ever — 'I shall die some day', says you.

It is interesting to note that A. B. 'Banjo' Paterson expressed a similar sentiment in his 1902 poem 'It's Grand', as commemorated in the Writers Walk on the eastern side of Sydney Cove (once part of the Phillip Domain):

It's grand to be an unemployed
And lie in the Domain,
And wake up every second day —
And go to sleep again.

Poor old Henry. When he was alive they hardly wanted to know him, and he had to cadge his drinking money. Then when he died they all wanted to claim him for their own — as a 'mate' for tourism or commerce — and he's even been used to sell margarine! The following poem, dedicated to the Henry Lawson statue in the Domain, was in part a trying to reflect this complex of emotions and the great irony that the man who was always broke should be featured on the old ten-dollar note.

The Man on the Ten-Dollar Note

The burghers of one-horse hamlets who now claim
Him on billboard or hoarding, tinsel and glitter:

The biggest/oldest/deepest something in the
world,
Cheap handouts at petrol stations for a full tank:
Didn't spare a quid for the man on the ten-dollar
note,
On the long haul from digger's tent to a fame
of sorts.

A young man with coal eyes who couldn't spell,
Wrote sad/funny yarns of black-stump and
drought;
Hoar-frost and stringybark, and of squabbling
mates,
And swagmen welcome as farmer's friends: for
all the
Gold in them thar hills, there's a bloody lot of
dross:
Kids stone an old drunk coming home in the
dark.

A farewell to stairwells and the clacking tram,
Red Page and Marble Bar; a simple monument
beside
The sea, the scattered hope of fragrant foreign
weeds;
The big-smoke buggers-up as well as hay seed
can:
Under a cabbage-gum on the road to Yurong
Point,
Australian Poet/Story without end/Pass round the
Hat.[35]

Mrs Macquaries Chair

Location: Mrs Macquaries Point

At the northernmost point of Mrs Macquaries Road (see Chapter 2) overlooking the harbour (Mrs Macquaries Point), a seat was carved out of the rock ledge, commanding a fine view. Mrs Macquarie was very popular. The seat became known as Mrs Macquaries Chair and remains a notable Sydney landmark.

Even today, many Sydneysiders still refer to 'Lady' Macquaries Road/Chair/or Point, as though her husband had been knighted. Philip Geeves made the point that such comments are part of our invisible cultural baggage from the convict days.[36] Elizabeth Macquarie was the Governor's lady — so it would have taken a headstrong purist or exclusionist to correct the convict or emancipist who spoke of her with such veneration — and old traditions die very hard indeed.

Further south along the existing road at the eastern side at the point, an additional inscription is carved into an outcrop of sandstone rocks, which reads: 'Mrs Macquarie's road 1816'.

Mobius Sea

Location: Crescent Precinct of Domain near Art Gallery of New South Wales

This statue by Richard Goodwin, winner of the Royal Blind Society's Sculpture Award (sponsored by James Hardie Industries) was unveiled by The Hon. Neville Wran, Premier of New South Wales on 1 May 1986. The statue, which originally stood in the Domain in Macquarie Street (opposite Bridge Street) was moved to its present site in the Domain near the Art Gallery of New South Wales.

Morshead Fountain

Location: In Domain opposite State Library

The Morshead Fountain of New South Wales, at the Morshead Gate of the Royal Botanic Gardens, was unveiled by Sir Roden Cutler in 1966 in memory of Lieutenant-General Sir Leslie Morshead and all who served with him, 1914–18 and 1939–45.[37] Morshead was awarded a knighthood for his defence of Tobruk in World War II with the 9th Division.

The fountain was designed by the architects Woodward, Taranto and Wallace. The fountain head was replaced with a new design by Phil Taranto in 1983. The new fountain comprises a central

> column composed of 36 evenly spaced steel rods, each ending in a petal motif, and enclosing a central pillar covered with orange tiles, with water jetting up inside the column and flowing over the petals to drop into the pool below. This pool is framed by metal letters on a circular pebblescrete surround, itself enclosed by a border of differently coloured pebblecrete from which rises a series of truncated pillars. Plaques are mounted vertically and horizontally on one of the truncated pillars. The central tiled column is surmounted by a cast bronze sculptural representation of the Rising Sun badge and has on its face seven colour patches of AMF units.[38]

Henry Moore, 1989, *Jaime Plaza.*

Henry Moore Reclining Figure — Angles

Location: Crescent Precinct Domain, near Art Gallery of New South Wales
The reclining bronze figure, 113 × 220 × 157 centimetres, with a green patina on a bronze base 11 centimetres high, was produced by the English sculptor, Henry Moore (1898–1986) — one of the greatest of twentieth-century sculptors. The inscription on the top of the base reads 'Moore 1/9' with 'H. Noace, Berlin' inscribed on the side. The original plaster was made in 1975, and cast in an edition of small bronzes in 1978. This cast, purchased from the artist in 1981, was the first of a proposed series of nine large casts.[39]

The Offerings of Peace and War

Location: In front of the Art Gallery of New South Wales
These two massive sculptures — the Offerings of Peace and the Offerings of War — are both the work of the English sculptor, Gilbert Bayes (1872–1953). They were commissioned in 1916 in the middle of the Great War (World War I) from 18-inch bronzed plaster models submitted to Sydney in 1915. They were cast at the foundry of A. B. Burton, Thomas Ditton, London, in 1923, and erected in front of the Art Gallery of New South Wales in 1926. War was exhibited in the courtyard of the Royal Academy in 1918 and attracted much attention.[40]

Both statues are of a man sitting on a horse — War with helmet and lance or spear, Peace with his olive branch and paradoxically a stocking full of heads.

The inscriptions on the statues are, for Peace, 'THEIR REAL AND LASTING VICTORIES ARE THOSE OF PEACE AND NOT OF WAR'; and for War, 'AND THAT JUSTICE AND MERCY GROW THAT OUR HOUSE MAY STAND FOREVER'.

Plaque to Anzac Buffet

Location: On the rock near the ladies toilet in the Phillip Precinct of the Domain
The plaque to the Anzac Buffet, reads:

> This spot marks the position of The Anzac Buffet, where, during the years 1916 to 1920, 82,000 soldiers and sailors were officially welcomed back, on their return from The Great War.
>
> Erected by the members of the Mosman No. 8 Red Cross Voluntary Aids.

The Anzac Buffet was an enclosed area and a building in the Domain 'where wounded and sick men [from the Great War were] met, on arrival in Sydney, by their relations and friends, and attended to by the Red Cross Voluntary Aids, before going home'.[41] In 1917 the building originally erected at the entrance was removed, and replaced by a suitable gate.[42]

> Huge numbers of returned soldiers were welcomed home [in 1919] by relatives and friends who assembled at the Buffet on 'troop' days . . . On one day 1,600 soldiers were given a hearty welcome by about 6,000 people, who were [all] admitted to the grounds of the Buffet during the day.[43]

Over the following years many trees and shrubs were sent from the Gardens to municipal and shire councils to plant avenues to soldiers who had died in the war.[44]

Plaques to Sudan Contingent

Location: Tarpeian Wall, opposite Sydney Opera House

There are two plaques to the Sudan Contingent that left Sydney on 3 March 1885, the first being inscribed:

> Sudan Contingent
>
> The New South Wales Contingent for the Sudan War (consisting of some 750 men and 225 horses) embarked on the troopships *Australasian* and *Iberia* at the wharves on the northeast side of the Circular Quay on 3 March 1885.
>
> The Contingent returned to Sydney in the troopship *Arab* on 23 June 1885.

This memorial donated by Thomas Gunting esq. RAHS.

The second plaque, to commemorate the centenary of the original departure, reads:

> To commemorate the Centenary of the departure of the NSW Contingent for the Sudan on 3rd March, 1885 His Excellency Air Marshal Sir James Rowland KBE, DFC, AFC, unveiled this plaque on the 3rd of March, 1985 following a parade from Victoria Barracks Paddington . . . of units of the Royal Australian Navy, The Australian regular Army and the Army Reserve, successors of the original Contingent.

Sir John Robertson, 1989, *Jaime Plaza.*

Sir John Robertson Statue

Location: Phillip Precinct of Domain

The bronze life-sized statue, erected to the memory of Sir John Robertson, on a trachyte pedestal with rock-faced shaft and classical leaf cornice capping, was unveiled by the Governor, Admiral Sir Harry Holdsworth Rawson, on 24 September 1904.

Sir John's father, James, had arrived in Sydney with his family in 1816 to work as a serviceman at the Parramatta Observatory erected by Governor Sir Thomas Brisbane. Sir John Robertson was an enterprising politician who was five times Premier of New South Wales. He died in 1891 and was given a state funeral. The statue was paid for by public subscription.[45]

It is appropriate that his statue stand in the Domain because from 1860, the Domain, under the administration of then Lands Minister Robertson, remained open at night to pedestrian traffic, allowing the public to use this valuable recreational resource during summer evenings. Thus the Domain was to become 'The Park where the Gates Never Close', although carriage traffic remained restricted after dusk for many years.[46]

The campaign to raise funds for the Robertson statue was mounted by *The Dawn* (a magazine for women edited by Louisa Lawson, mother of Henry Lawson) in recognition of his work for free settlement and education.[47]

Originally the statue was on the lawn not far from the principal entrance (Morshead Gate) facing the Public Library in Macquarie Street.[48] It was moved in 1912, during the widening of Macquarie Street during the minor re-location of the Palace Garden Gate. The statue, now isolated, was moved to 'the north end of the shrubbery border which immediately faces the visitor on the western side of the asphalt walk when he enters the Grounds at their principal entrance' (as seen in the 1914 and 1921 maps) and its bareness (was) 'relieved by a background of trees and shrubs which are growing in the border'.[49]

On 28 April 1982, I had a phone call from Mr N. K. Lamport of Pymble, a relative of John Robertson, who confirmed the statue was once 'in the open space opposite the Mitchell Library looking at the city'. He also reported that at one stage the statue had 'a hive of bees in its ear', and was moved again, 'but not on this account but because of the Cahill Expressway, to occupy its present site in the Phillip Precinct of the Domain.'[50] I'm not so sure about the bees as he doesn't appear to have an earhole!

Shakespeare Memorial (not on Trust Land)

Location: Traffic Island, Cahill Expressway, opposite Mitchell Library and State Library of New South Wales The Shakespeare Memorial by Sir Bertram Mackenal, erected near what was then the Mitchell Library on 3 February 1926, formed a portion of a city plaza being created between the Library and the Botanic Gardens, with shrubbery and trees as a background. According to an ink drawing by Lloyd Rees in 1926, the statue was then aligned to face the Palace Garden Gates prior to their being moved to Macquarie Street (see p.40). The current site of the memorial is near where the Bourke statue originally stood (see p.109).

The memorial consists of a massive marble pedestal with grouped figures (each about 2.5 metres in height) of some of Shakespeare's more famous characters — Romeo and Juliet, Falstaff, Hamlet, and Portia — surmounted by a 2.75 metre statue of the poet and playwright. The bronze figures were cast in England; the marble was burnished in Italy.

The upper portion of the pedestal contains the following inscription:

> Shakespeare
> 1564–1616
> Presented to the City of Sydney by
> Henry Gullett
> August, 1914

On the base of the column there is the following wonderful quotation from The Tempest:

> Our Revels now are ended. These our actors
> As I foretold you, were all spirits and
> Are melted into air, into thin air:
> And like the baseless fabric of this vision,
> The cloud-capp'd towers, the gorgeous palaces,
> The solemn temples, the great globe itself.
> Yea, all which it inherit, shall dissolve.
> And like this insubtantial pageant faded,
> Leave not a rack behind. We are such stuff
> As dreams are made on, and our little life
> is rounded with a sleep.

The Six Day War, 1989, *Jaime Plaza.*

Sphinx from estate of Sir James Martin, 1989, *Jaime Plaza.*

On 23 April 1926 a ceremony was held at the Monument to honour Shakespeare's birthday. The large gathering was addressed by Sir Mark Sheldon and wreaths were laid on the statue by members of the Shakespeare Society, the English Speaking Union, and students from the Presbyterian Ladies College and Osborne College, Blackheath.

(The) Six-Day War

Location: Crescent Precinct of Domain near the Art Gallery of New South Wales, installed *c.* 1984
This 225-centimetre steel sculpture, entitled The Six Day War II (1973) by Margel Hinder, born 1906, was purchased from the artist by the Art Gallery of New South Wales in 1980 for $16,000.[51]

Sphinxes

Location: Crescent Precinct of the Domain near swings
Two sandstone sphinxes also from the estate of Sir James Martin (see Choragic Monument) moved to the Domain *c.* 1943.

Chronological Checklist of Statues, Plaques, Fountains and Memorials in the Domain

1816	Mrs Macquaries Chair
1842	Governor Bourke Statue
1884	Drinking Fountain, southern end Art Gallery Road (not on Trust Land)
1885	Plaque to Sudan Contingent
1904	Sir John Robertson Statue
1905	Robert Burns Statue
1908	Fleet Steps
1923	King Edward VII Statue
1925	Matthew Flinders
1926	Shakespeare Memorial
1926	The Offerings of Peace and War
c. 1920s	Plaque to ANZAC Buffet
1931	Henry Lawson Statue
1943	Sphinxes (from estate Sir James Martin)
1950	Desert Corps Horse Memorial
1965	Bronze Horses (miniature pair)
1966	Morshead Fountain
1981	Henry Moore Reclining Figure — Angles
1984	(The) Six-Day War
1985	Brass Plaque to Flinders' Cat Trim
1985	Centenary Plaque to Sudan Contingent
1986	Mobius Sea
1991	Almost Once

7

Dearly Departed

Alphabetical Checklist of Statues, Plaques, Fountains, Memorials, etc. No Longer in Gardens and Domain

Albert the Good (Victoria and Albert)

Source: William Theed, the Younger (British) (1804–91)
Date/Place: 1866
Removed: 1987
Additional Information: A bronze statue of the young Queen Victoria had been designed for Chancery Square (now Queen's Square) and the foundation stone for the statue had been laid in Chancery Square by two grandsons of Queen Victoria, the two princes (see p.54), during their visit to Sydney on 6 August 1881. The day was declared a public holiday so as many people as possible, and especially schoolchildren, could watch the princes lay the stone.[1] But the statue was temporarily displayed in the Garden Palace Exhibition in the Gardens and was destroyed when the Garden Palace burned down on 22 September 1882.[2]

After considerable agonizing, the Agent-General in London commissioned Joseph Boehm to produce a replacement statue for £3,000. The contract stipulated that the statue should be unique and no copies be made, so there were considerable bureaucratic flutters in Sydney towards the end of 1887, when it was discovered that a replica of the Boehm statue had been presented to the Queen by her son and heir and installed atBalmoral. It transpired that the prince had won the statue in a game of cards; there was no legal action because of the royal connection.[3]

Meanwhile the overlife bronze statue of Albert, consort to Queen Victoria, by the English sculptor William Theed, the Younger originally stood at the northern end of Hyde Park, and was unveiled by Sir John Young in April 1866.[4] William Theed did many works for the Prince Consort and had been chosen to take his death mask by Queen Victoria in 1861.[5]

The statue of Albert must have been a point of social focus at the time. Joy Thompson has a letter from her grandfather as a child, dated 16 April 1866, in which he states, 'All the National Schools are going to walk on St. George's day to Prince Albert's statue'.[6]

A Prince Consort's Statue Association had been formed with headquarters at Thomas Mort's Building. Subscribers donations were recorded in the *Sydney Morning Herald*. The sculpture shows Albert in the robes of Knight of the Garter, with breeches, slippers and a full-length robe held by gold chains. In his left hand he is holding a plan of the Crystal Palace, while leaning on a column embellished with a portrait of the young queen, a garland, and a coat of arms.[7]

When Albert (born 1819) died prematurely in 1861, aged 42, Victoria was devastated and suffered a nervous breakdown, and as part of the official mourning the wooden edges of the showcases in the Victoria and Albert Museum were painted black. Museums all around the then British Empire followed suit — a tradition that lingered in Australian Museums until the 1970s! To the end of her long life Victoria slept with a photograph of Albert above her bed, taken of his head and shoulders as he lay dead. Every night at Windsor, his clothes were laid out on the bed, and every morning the basin in his room was filled with fresh water.[8] And because Victorians were such a death-oriented culture, their statues became their memorials.

The Sydney (Boehm) statue of the older Queen Victoria, holding sceptre and orb, and mounted on its

Albert the Good in original location in Hyde Park, *David Churches.*

great pedestal of Moruya granite, was unveiled by Lady Carrington, wife of the Governor, on 24 January 1888, to celebrate 100 years of European settlement.[9]

In August 1922 the statue of Albert was moved from Hyde Park to what was the Inner Domain, opposite the Chief Secretary's building, because of city railway extensions. Victoria was later moved back from the mainstream of traffic and set so she was able to see St James' Church.[10]

In the 1950s Albert stood on his massive granite pedestal and sandstone plinth in the Gardens, looking north, and about in the same north/south axis as King Edward VII.[11] He was then moved again during the construction of the Cahill Expressway, to a site in Lawn 15, under a fig tree in the Palace Garden, just north of where the New Rose Garden was to be constructed.

Meanwhile the eastern end of King Street between St James' Church and the Law Courts Building, containing the Supreme Court of New South Wales, was changed into a pedestrian plaza.

Finally, as part of the bicentennial renovations of Macquarie Street, the peripatetic statue of Albert was to move in 1987, back to the widened eastern footpath near the Hyde Park Barracks, to be near his queen once more.[12] A noticeably younger Albert now gazes towards an older Queen but she, now facing north, cannot gaze at him — for when she was moved to the southern end of the plaza there was considerable, and not totally facetious, concern 'that she not turn her back on the law' (as apparently she had in the 1960s).

Andromeda

Source: James Forsyth (British; 1826–1910) — inscribed 'James Forsyth Sculp. London'
Date/Place: Purchased 1892, gift to Gardens 1958
Removed: No trace in 1972. Destroyed by vandals?
Additional Information: Donated by Art Gallery of New South Wales, (4630), purchased Sydney, 1892 — 44-inch marble, the original of 20 copies.

Apollo Belvedere

Source: —
Date/Place: 1903, Palace Garden
Removed: Presented to Sydney Boys High School, 1945, by the Hon. R. J. Heffron. Sydney Boys High School Speech Day Publication, 1945.
Additional Information: Copy of a classical marble with right hand resting on a snake-entwined support and his extended left hand holding a plectrum.

Apollino

Source: —
Date/Place: 1903, 'at corner of rockery on eastern side of creek near kiosk'. Bed 14 in 1965
Removed: In store in 1972
Additional Information: In storage — broken left arm above wrist (1988). This may or may not have been the 'Apollo' reported as being destroyed by Gardens employees by Mr Kirkpatrick.[13]

Architecture

Source: —
Date/Place: 1903, Palace Garden. 1914, Lawn 3 Upper Garden
Removed: —
Additional Information: Small white marble statue of Architecture.

La Ballarina

Source: Italy 1883
Date/Place: 1903, Lower Garden near Sundial 1903. 1965, near Bed 98
Removed: 1972
Additional Information: 'A girl pausing amid the dance, and bearing in one arm a wreath of flowers.' Vandalised, in storage (1988).

The Bather

Source: Italy 1883
Date/Place: 1883, Lower Garden. 1965, Bed 121
Removed: There in 1965
Additional Information: 'A girl with a close-fitting tunic, just sliding down into the water.' Vandalised (decapitated), in storage (1988).

(Gardens) Bell

Source: —
Date/Place: Dated 1830, situated between Aviary (site of present Succulent Garden) and Office
Removed: 1901, relocated at State Nursery, Campbelltown — returned to Gardens in 1931
Additional Information: A bell, dated 1830, was used to ring out the hours from 1840 to 1901. It rang four to six times daily, according to the season, and hung between the Aviary (site of the present Succulent Gardens) and the Office (site of the present Visitor Centre). In 1901 the Garden Bell ('which was not in a central position for the whole of the staff' and 'which had proved too small in these noisier times') was relocated at the State Nursery at Campbelltown, and 'a larger bell was installed in the work-yard [Depot?] which is nearly in the centre of the grounds' on 13 October 1901.[14] The Campbelltown Nursery was dismantled in 1931 and the bell returned to the Botanic Gardens.[15] This bell was then rung at closing times in the Gardens — a practice which continued into the late 1950s and possibly the early 1960s.[16]

Bird Fountain

Source: Woman's Christian Temperance Union
Date/Place: 28 September 1939, in region of what is now the Succulent Garden
Removed: 1960
Additional Information: Frances Willard Memorial Pond. Brass plate there in 1963. The memorial consisted of a circular pool surrounded by a stone coping with a small central spherical ornament which was removed *c.* 1960 due to erosion.

Blind Nydia

Source: G. Andreoni, nineteenth-century Italian
Date/Place: Transferred from Art Gallery of New South Wales in 1958
Removed: Destroyed by vandals — no trace in 1972
Additional Information: Presented to the Art Gallery of New South Wales by A. Stranger Leathes in 1889 — 48-inch marble.[17]

Boer Gun

Source: Boer War
Date/Place: 1906, Palace Garden
Removed: —
Additional Information: A gun captured during the Boer War was placed in the grounds near the Main Entrance (Morshead Gate), not far from the then position of Sir John Robertson's statue.[18]

Canoe

Source: Pitcairn Island Relic
Date/Place: Brought here when inhabitants of Pitcairn Island moved to Norfolk Island, in 1856
Removed: *c.* 1860
Additional Information: 'A canoe [associated] with the Mutiny of the Bounty is rotting in an unfrequented corner of the Gardens. It much resembles the canoe so popularly represented as having been used by the famous Robinson Crusoe.'[19]

Statues along Main Walk in Upper Garden, *RBG Library*.

Ceres

Source: —
Date/Place: 1914, Upper Garden
Removed: —
Additional Information: This statue is shown in the 1914 map in the region of theUpper Garden (about where the Access Road now runs).

Commerce

Source: —
Date/Place: 1903, Palace Garden. 1914, Upper Garden
Removed: —
Additional Information: Small marble statue situated in what is now the corner of Bed 13, east of the pathway heading towards the Wishing Tree. Previously in the Palace Garden, removed 1921. In 1914 there was another statue of Commerce near the Levy Fountain, still there in 1921.

Diana

Source: Achille Simonetti, sculptor
Date/Place: 1890s, Lower Garden area
Removed: —
Additional Information: Marble by Achille Simonetti, model for arms and legs — Hannah Fitzsimons.[20]

Robert Dawson Memorial

Source: —
Date/Place: Mentioned in 1917 in Mitchell Library in letter from R. L. Dawson to a friend
Removed: 1950s
Additional Information: Plaque on stone plinth to Robert Dawson, involved in diverting water from Hyde Park to Botanic Gardens Creek in 1850s. Removed from Gardens in 1950s.

A 3-inch pipe from Busby's Bore in Oxford Street,

laid in 1903 as far as the Botanic Gardens gates but never used, was continued in 1907 to the head of Garden Creek, at Fig-Tree Bridge, at a cost of £27 17*s*. 3*d*. This water, used for flushing the creek, was not considered suitable for drinking.[21]

From the time of foundation, the question of a water supply to the Botanic Gardens was a serious problem. In the early days the Botanic Gardens Creek, 'a beautiful stream of fresh water, with pretty falls here and there',[22] was used but often failed, as its catchment extended only to the vicinity of St Marys Cathedral. As there was no well in the Gardens the next supply of water came from the swamp near the cricket ground (Hyde Park) and then from Busby's Bore (Waterloo).

John Busby and his son William oversaw the construction of a tunnel by convicts from a swamp at Waterloo, started in 1827 and finished in 1837. The bore began supplying water by seepage in 1830, and Busby's Bore, as it was known, served Sydney with water until 1858. A cairn to Busby's Bore was unveiled in Centennial Park in October 1988.

A proposed reservoir in Hyde Park was never constructed, the water being supplied through a standpipe.[23] After the completion of Busby's Bore the Gardens were supplied with water in times of drought by carting water in a barrel from the pipe head in Hyde Park.[24]

Allan Cunningham, writing in 1837, had drawn attention to the inadequate water supply.[25] He recommended that the Garden be furnished with water from the Tunnel in Hyde Park (Busby's Bore), which had already been conveyed to the convicts' barracks, and might be easily conducted thence by a line of pipes through the Upper Domain to a tank or reservoir within the Garden.

In 1854 when Sydney's water supply from Botany Swamps appeared to be failing, Robert Dawson proposed an engineering solution to increase the flow by pumping, and helped extend the life of the system until alternatives could be found.

On information provided by Mr A. Greenhalgh, there was once a plaque to Robert Dawson on a stone plinth on the eastern side of the stream south of the Spring Walk. Mr Greenhalgh told me that Robert Dawson was the first manager of the Australian Agricultural Company and came initially to Australia in 1825.

The plaque was removed in about the 1950s. It may have been that the construction of the Cahill Expressway and its subsequent destruction of the flow of water into the creek, rendered the sign obsolete.

In 1903 Busby's Bore was continued into the Domain and the water ran into the head of Botanic Gardens Creek.[26]

Discobolus (Discus thrower)

Source: —

Date/Place: Originally in Bed 52, Lower Garden below Refreshment Room, relocated adjacent to terraced walk above the Old Rose Garden near Darnell Smith Memorial Seat. Probably installed in 1879, leading up to the International Exhibition

Removed: Destroyed by vandals early 1970s

Additional Information: The original bronze, by the Greek sculptor Myron, now in the British Museum, formed a group with Athena and stood on the Acropolis, about 460–50 BC. A marble copy was made and is now in Rome. The marble copy in the Gardens was probably made in Italy. As of 1988, only its feet were left in storage.

Drinking Fountains

Last century drinking fountains were literally that — monumental fountains with a series of small spouts, often equipped with drinking cups on chains. The position of drinking fountains in the Gardens were marked on Maiden's 1903 map as small squares with a dot in them. J. H. Maiden, in his evidence to the Royal Commission on the improvements of Sydney, 1909, said:

> Every street in Sydney should have its drinking-fountain. It should be at least as easy to obtain a drink of pure water as a glass of milk or beer. Sydney is a sub-tropical city, and sometimes it is warm and sometimes it is dusty . . . Sydney should be a city of fountains.[27]

In Maiden's time there were many drinking fountains in the Gardens:

(a) Palm Grove

Source: —

Date/Place: Located at entrance to Palm Grove, marked as K on Maiden's map of 1903

Removed: *c*. 1917

Additional Information: The ornate marble drinking fountain in the Palm Grove (site identified by old

Marble Drinking Fountain in Palm Grove, 1887, *RBG Library*.

photographs) was surmounted by small marble(?) statue of girl (angel?) on a stand supported by four columns. Maiden complained that: 'certain mischievous boys and other people delight in wrenching off the cups, damaging the spouts and injuring the fountains in various ways'. By 1914:

> The old taps and enamel mugs, which had been in use for drinking purposes for varying periods . . . [were replaced with] . . . bubbling fountains . . . which have been in use in the Botanic Gardens and Outer Domain for some years. They are more sanitary, and are appreciated by the public.[28]

The existing sandstone bubblers in the Gardens probably come from this era. The new bubblers, on the path from the Tropical Centre to the Palm Grove and at the top of the Palace Garden Steps, were made in the same style.

(b) Near Victoria Lodge and Allan Cunningham Monument
Source: —
Date/Place: 1863 near Victoria Lodge and Allan Cunningham Monument
Removed: 1903
Additional Information: Two cement and concrete (assumed drinking) fountains were erected in the Lower Garden. They had become 'insanitary and beyond repair' in 1903 and were removed.[29] A new drinking fountain, imported from Glasgow at a cost of £47 10*s*. 0*d*., erected on a suitable and convenient corner of ground at Victoria Lodge, directly opposite the Garden gates — northern entrance. Cost of circular stone steps of hard freestone and fixing same, £20. Plumbers' work — providing all necessary piping and connections, new taps, and laying-on water, £20.

(c) Main Gate, Fig Tree Avenue
Source: —
Date/Place: 1866
Removed: Destroyed during construction of Cahill Expressway
Additional Information: Terracotta and stone drinking fountain at end of Fig Tree Avenue, outside the main gate to the Upper Garden, of what is now the Phillip Precinct of the Domain. This was the only drinking fountain in the Domain and it was 'improved' in 1897 so that stock would not drink from the same basin.

Flora

Source: Sculptor William Lorando Jones, born Merthyr Tydfil, Wales *c.* 1820, died *c.* 1890
Date/Place: 1870s, Upper Garden (about where Access Road is now). Somewhere near Beds 70–71 in 1903

Drinking Fountain in Domain at end of Fig Tree Avenue near Main Gate, *c.* 1870, *David Churches.*

Removed: —
Additional Information: William Jones came to Australia *c.* 1856 and later moved to Granville, N.S.W., and showed eight works at the Intercolonial Exhibition, Sydney, 1870. His career was ruined because of his known atheism, and in 1871 he was sentenced to two years (reduced to one month) for blasphemy. Later, in the 1870s, through the influence of Sir Henry Parkes, his statue Flora was placed in these Gardens.[30]

Fort Macquarie

Source: Guns from HMS *Supply*
Date/Place: 1789 site of Opera House
Removed: Prior to construction of tram shed *c.* 1902
Additional Information: In July 1788, Lieutenant Dawes of the Marines, was directed to construct a small redoubt of the east side of Sydney Cove, which was finished in November 1788. In 1791 this redoubt was demolished and the guns and flagstaff moved to the west side of Sydney Cove, and a storehouse was built on the site of the redoubt.[31]

A half-moon battery at the extreme end of Bennelong Point was completed in December 1798, and a few guns from the *Supply* mounted by the ship's company. At the foot of the cannon were large flat stones on which the Aborigines baked their fish whole.[32] This was later expanded and became known as Fort Macquarie.

Fort (Mrs Macquaries Point)

Source: Cannon sent from Britain following report of Select Committee in 1853
Date/Place: 1856 Mrs Macquaries Point
Removed: 1870
Additional Information: Mrs Macquaries Point was once the location of Macquarie Point Battery (1856–70), immediately behind Mrs Macquaries Chair,[33] one of the early artillery fortification of Sydney, not to be confused with Fort Macquarie on Bennelong Point.[34] 'Lady [*sic*] Macquarie Battery' was the scene of a 'moonlight sham fight' as reported in the *Illustrated Sydney* News of 16 May 1867.[35] There was another sham naval attack on Mrs Macquaries Point on 12 April 1881.[36]

(Ornamental) Fountains

In 1867 (before Sydney became the 'big smoke') Leigh & Company, reporting on the Gardens, said: 'The beautiful fountains give a cool and refreshing appearance to . . . the gardens: the ponds which surround them are filled with a number of gold carp and other fish'.[37]

The water pressure was always a problem in the early days and Maiden, reporting to the Royal Commission in 1909, said he hoped the fountains would play more vigorously when the Cataract Dam is in full action. Pollution had become a problem even then, when 'the smoke from certain huge belching chimneys [filled] the water-basins with large and numerous smuts, which . . . destroyed all the lilies and gold-fish . . . put into them. And filthy, scummy water is not a beautiful object to contemplate.'[38]

Ornamental fountains were found in the following sites in the Gardens:

(a) (Site of) Choragic Monument (Iron Fountain)
Source: —
Date/Place: Prior to 1892, site of Choragic Monument
Removed: 1943
Additional Information: Circular pond and fountain, in Lower Garden. Replaced by Choragic Monument in 1943. In 1903 Maiden refers to a handsome iron fountain, with flower beds around the large cemented basin which encircled it.[39]

(b) Palace Garden Fountains
Source: —
Date/Place: 1894, Palace Garden
Removed: One fountain near Morshead Gate was removed in 1899; the northernmost one in 1921; the remaining two were on the 1927 plan, but had been removed by 1954
Additional Information: The 1894 Survey of the Gardens shows four small fountains in the Palace Garden. The erection of the Governor Phillip statue rendered the small basin fountain at the principal entrance unnecessary, owing to the larger fountains of the statue — about 100 yards away — dwarfing the smaller one. Furthermore, being so near the road, it was a standing temptation to children to throw rubbish into it. Its removal improved the appearance of the entrance, which formerly had a rather cramped appearance. The site was ballasted and gravelled in the usual manner.[40]

(c) Fountain for Crippled Children
Source: Designed by Peter Spooner, Associate Professor of Architecture, University of New South Wales

Fountain in Lower Garden (site of Choragic Monument) with Band Rotunda in background, *David Churches*.

Date/Place: Macquarie Street entrance to Palace Garden Gate *c.* 1966
Removed: 1987 in association with bicentennial renovations of Macquarie Street — and replaced with three new fountains in this area
Additional Information: The 'forget-me-not' Wishing Well at the Macquarie Street entrance to the Palace Garden Gate was designed to raise funds for the New South Wales Society for Crippled Children.[41]

(d) Fountains at Aviary Site
Source: —
Date/Place: The Aviary, 1861
Removed: —
Additional Information: S. T. Gill's 1861 drawing of the Aviary in the Gardens shows two symmetrical ornamental fountains at each end of the aviary.[42]

Grave of Joseph Gerrald

Source: Joseph Gerrald, died 1796, aged 36
Date/Place: Middle Garden — halfway between the creek and what was to become the Wishing Tree
Removed: Stone removed *c.* 1807
Additional Information: In 1794, the Scottish Martyrs were transported for comparatively mild political dissent.

In the years following the French Revolution and the publication of the *Rights of Man* by the American revolutionary Thomas Paine, people in England, Scotland and Ireland advocated parliamentary and constitutional reform. The Tory government, over-reacting to the democratic movement, tried to make an example of its leading proponents by convicting them of 'sedition' and transporting them to New South Wales as convicts.

One of the martyrs was Joseph Gerrald, a young barrister who was sentenced to 14 years for sedition for seeking to have annual (regularly elected) parliaments. Born in the West Indies of an Irish father and English mother, Gerrald was educated in England and had practised at the bar in America. He was one of the most brilliant of his band of martyrs, warm and impetuous — an eloquent speaker and reformer — and therefore dangerous. In an age when few had the vote Gerrald was seeking to make Parliament more accountable, and having observed the more democratic election system in Philadelphia he noted that, 'Parliaments have generally been dangerous in proportion as they have been long'.

Gerrald was finally brought to trial on four accounts: Making use of French terms (words); specifically having used the word 'citizen'; attending illegal meetings, and having urged arguments in support of annual parliaments.[43] As he was talking about liberty and individual freedoms, he was charged with sedition. Freedom was a dangerous idea, and 'citizen' a dangerous word, for hadn't the French revolutionaries written a song that urged their citizens to take up arms and to form their battalions? To read his spirited self-defence almost 200 years later made my spine tingle — if only as an affirmation of the power of the printed word.[44]

In the colony the martyrs were treated as special prisoners. Gerrald, his health declining, purchased and occupied the house on the site of the Old Farm and worked the land until his untimely death a few months later of consumption in 1796, aged 36. Before he died he said, 'I die in the best of causes: and I die, as you are my witness, without repining'.

Joseph Gerrald was buried at Farm Cove and on his tombstone was recorded: 'He died a martyr to the liberties of his country'. The precise location of his grave has now been forgotten, but legend has it he was buried half-way between the creek and what was to become the site of The Wishing Tree.

Hero

Source: G. G. Fontana, Italian sculptor
Date/Place: Gift to Gardens, 1958
Removed: Destroyed by vandals
Additional Information: Donated by Art Gallery of New South Wales, 51-inch marble

Industry

Source: —
Date/Place: 1903, Palace Garden. 1914 Lawn 1, Upper Garden
Removed: 1905(?)
Additional Information: This small white statue was thrown bodily from its pedestal in 1905 and very much injured, thus necessitating its repair.[45]

Iron Church

Source: Brought from Glasgow in 1855
Date/Place: 1855, site of extensions to State Library of New South Wales, facing Macquarie Street
Removed: 1879
Additional Information: An Iron Church (Presbyterian Church of St Stephen's) was transported from Glasgow in 1855 and erected on the site of the extensions to the State Library of New South Wales, facing Macquarie Street. In 1874 it housed the lending branch of the Public Library. It was moved to Lidcombe in 1879, and demolished in 1958.

Macquarie Point Battery
(see Fort, Mrs Macquaries Point)

(Marble) Lions

Source: Italian marble on sandstone base
Date/Place: *c.* 1882 Palace Garden
Removed: Taken to Centennial Park during construction of Cahill Expressway
Additional Information: 'On either side of the main walk is a fine sculptured lion in white Italian marble resting on a sandstone base.'[46] The lions were originally positioned during the laying out of the Garden Palace Grounds. In 1909 the pedestals of these lions had settled down from the perpendicular.[47]

They were removed from the Gardens during the construction of the Cahill Expressway and taken to Centennial Park. Currently in storage (1988). The Director is considering replacing them in the Gardens, near the Rose Garden Gate, looking towards the Conservatorium of Music.

(Small) Lion

Source: —
Date/Place: Region of Aviary (site of present Succulent Garden) in 1861. Still there in 1925
Removed: —
Additional Information: Joan Moore (of the Friends of the Gardens) showed me a photograph of herself, aged about four years, beside this lion in 1924/5. This small lion, (of unknown material) with its head turned to one side, was unlike any other lion in the Gardens that I had seen to that time. Joan Moore said her family used to visit the Gardens from Balmain by public transport. She

Marbles Lions in Palace Garden, 1914, *RBG Library*.

identified this as being the lion in an etching of the Aviary in the Gardens by S. T. Gill, dated 1861.[48]

Mars

Source: —
Date/Place: 1914, Lawn 3 (Upper Garden)
Removed: —

Meleager

Source: —
Date/Place: 1903, Palace Garden
Removed: 1945
Additional Information: Copy of classical marble with dog and boar at his feet. Removed to Fort Street Boys High School (now Fort Street High School) in *c.* 1946 after storage in the Mitchell Library (during the war?).[49]

Mercury

Source: —
Date/Place: 1903, Palace Garden, between steps and Palm Grove, on right-hand side of pathway heading towards Palm Grove
Removed: —

Music

Source: —
Date/Place: 1903, Palace Garden
Removed: —
Additional Information: Small white marble statue.

Nymph (Oenone)

Source: Art Gallery of New South Wales, purchased 1887.[50] Inscribed W. Calder Marshall, RA
Date/Place: 1958, Middle Garden, later in Pond
Removed: Returned to Art Gallery of New South Wales 7 May 1973
Additional Information: 42½-inch marble nymph by British sculptor William Calder Marshall, RA (1813–94) — a nymph of Mount Ida, daughter of the River Cebrenum in Phrygia, married Paris, to whom she foretold the destruction of Troy. Removed from pond, cleaned of algae, and returned to Art Gallery of New South Wales in 1973. Two fingers missing — had been broken off years ago.

Nymph (Psyche)

Source: Presented to Art Gallery by Lady Frankfort, 1924[51]

Psyche by J.H. Dannecker, 1972, *RBG Library*.

Date/Place: Lower Garden facing the bridge at the northern end of the large pond[52]
Removed: 1973
Additional Information: 53-inch marble statuette by the German (neo-classical) sculptor J. H. (Johann Heinrich) von Dannecker (1753–1841). Returned to Art Gallery of New South Wales in 1973. The Dannecker sculpture in the Gardens was possibly the original as commissioned by 'the English General Murray in 1814' and finished in 1817.[53]

Nubian Figures

Source: —
Date/Place: Photograph 1909
Removed: —
Additional Information: Two statues of Nubian figures in the Palace Garden resulted in that area of planting (Bed 62 or Bed 72) being known for many years as the 'Black Boy' shrubbery. An iron negro figurine was moved at some time to Cottage 2 garden (Lion Gate Lodge); currently in Domain Depot, back compound.[54]

Queen Elizabeth II Walk

Source: —
Date/Place: March 1977
Removed: *c.* 1980
Additional Information: Circular Plaque on Gatepost, Queen Elizabeth II Gate, and another at Mrs Macquaries Point — dedicating Queen Elizabeth II Walk — now missing. Another circular bronze plaque in asphalt near Sea Wall signifying the spot where Queen Elizabeth II first set foot on Australian soil (asphalt) — disappeared *c.* 1983.

Rebecca at the Well

Source: Italy 1883
Date/Place: Bed 10, near Glasshouse No. 10, 1883–1965
Removed: There in 1965, removed *c.* 1970
Additional Information: 'A maiden bearing on her left shoulder a beautifully designed pitcher, and standing with uplifted drapery among a range of ferns.'[55] Damaged by a contractor's truck from Public Works Department and removed to Depot.

Retaliation

Source: 1958, Art Gallery of New South Wales, C. B. Birch. Inscribed C. B. Birch Sc. 1878
Date/Place: 1958–65 on main pathway through Palm Grove
Removed: There in 1965. Returned to Art Gallery of New South Wales 1977
Additional Information: 8 foot 4 inch inscribed bronze (3252) by British sculptor Charles Bell Birch (1832–93). Executed expressly for the Sydney International Exhibition of 1879 for which it was awarded a First Degree of Merit Special.[56] Purchased from the Art Gallery for £662 6*s.* 0*d.*[57] On 25 January 1961 an International truck, registration number FYT 603, owned by S. McCormack and Sons Pty Ltd, collided with the statue. Repairs to the marble base of the statue cost £152 7*s.* 6*d.* and were paid by the company.[58] Retaliation was returned to the Art Gallery (*c.* 1977) as a swap for The Satyr (see p.101) when they decided it should not have been released.

Science

Source: —
Date/Place: 1914, opposite Levy Fountain, Upper Garden
Removed: —

La Sonnambula

Source: Purchased by the Art Gallery 1880 in Sydney.[59] Giovanni Guiseppe Fontana (1821–93). Inscribed 'G. FONTANA Sc.'
Date/Place: Transferred from the Art Gallery of New South Wales in 1958. In second 'store' near staff buildings, 1972
Removed: Returned to the Art Gallery of New South Wales in 1972/73
Additional Information: 61-inch marble by the Italian-British sculptor Giovanni Guiseppe Fontana — awarded First Degree of Merit Special at the 1879 Exhibition. Returned to Art Gallery of New South Wales in 1972/3.

Sphinxes

Source: From estate of Sir James Martin
Date/Place: Bed 34 in 1903
Removed: To Crescent Precinct of Domain
Additional Information: From the estate of Sir James Martin, source of Choragic Monument (see p.88). From 1903 map from *Guide to Gardens* they were in Bed 9 (now Lawn 21) — they are now in the Crescent Precinct of the Domain near the swings (1988).

Spring

Source: Job Hanson 1879/Italy 1883/estate James Stedman
Date/Place: Palace Garden Steps/Bed 84/Near Lion Gate Lodge
Removed: See below
Additional Information: As noted in the *Illustrated Sydney News* of 25 October 1882, the statue of Spring imported by Job Hanson lost her arm during the Garden Palace fire, and showed the 'marks of repair' in 1903. The Charles F. Summers statue of Spring arrived the following year, as reported in the *Sydney Morning Herald*, 10 January 1883.

Statue of Spring prior to donation to the Gardens in 1957 from the estate of James Stedman. *Mervyn Doust.*

There were two statues of Spring in the Gardens when I first came to Sydney in the early 1960s — one in Bed 84 (the Charles F. Summers statue) and one under a Cedar of Lebanon at the eastern end of the Spring Walk (near Lion Gate Lodge). When I came to work at the Gardens in 1980, the latter statue, with her left arm broken off below the wrist, was languishing in the 'graveyard' behind the developing Succulent Garden. The Cedar of Lebanon was subsequently removed on 31 July 1984.

In 1984 the statue in Bed 84 was decapitated; the Ranger reported he had seen a party of drunken sailors on their way through the Gardens that night.

In 1987, the statue from the 'graveyard' that had stood near Lion Gate Lodge was restored, and placed on the Palace Garden Steps, along with the statue of Autumn from Bed 97, and two marble urns, with money provided from the Philip Geeves Memorial Trust. They were unveiled by their patron, Lady Fairfax, on 26 April 1988. Two Haddonstone urns, which had also been placed on the steps, were destroyed the night before the dedication.

Following a newspaper article in the *Sydney Morning Herald* the next day, with a photograph of the statue — its left arm now broken off below the elbow — I was contacted by Mr Mervyn Doust of Seaforth, who told me that this statue of Spring had belonged to his great-grandfather, the confectioner James Stedman, who imported it from Italy prior to 1880. The statue had been located in the rear garden of Stedman's residence in Cleveland Street, Redfern, from 1880 to 1890, then relocated to his new home at The Crescent, Manly, where it remained until World War I.

After Stedman's death, his eldest son, James Noble Stedman (grandfather of Mr Doust) transported the statue to Telopea Street, Wollstonecraft, about 1917, where it remained in the front garden until April 1957. James Noble Stedman died in 1948 but Mr Doust and his mother continued to live in the property until June 1957, when the home was sold and the statue donated to the Gardens.

Unfortunately this statue was again vandalised on the evening of 13 July 1989 and very badly damaged, as were the two marble urns. Following a newspaper report of this vandalism in the *Sydney Morning Herald* I had a letter from Judith Hunter about the Hanson statues.

Summer

Source: Job Hanson 1879/Italy 1883
Date/Place: Originally on steps to Garden Palace/Bed 53
Removed: One destroyed in 1948. The other in storage in 1972
Additional Information: The statue of Summer remaining on the pedestal at the top of the steps on the main walk to the Palace Garden Grounds was destroyed on the evening of 8 January 1948. The statue was knocked down the stairs and smashed beyond repair (badly damaging a step) and the head of the statue was removed. Another statue of Summer remained, the one referred to by Nan McDonald in her poem 'Wet Summer: Botanic Gardens':

Wet Summer: Botanic Gardens

Under the low dark sky and the sodden leaves
Poor Summer bared her shoulder with coy grace,
Her marble flesh streaked with mortality,
Her sheaf of wheat lay mildewed on her arm,
Her eyes stared vacant from her tear-strained face
On paths empty at noon, on glooms beneath
Great fig-trees, where the drop of rotting fruit
Broke the warm, damp hush, on unbelievable
green
Of wet grass still unfaded by the sun
In this strange season. Once we laughed to see
That foolish white thing named for the brazen
queen
Whose sword we knew, the fierce and splendid
one.

But tonight, alone with the steady sound of rain,
I do not smile, seeing her image there
On the haunted edge of sleep — the blackening
marble,
The blind and weeping eyes, the ruined grain —
Seeing this season of the world's despair.[60]

This statue, a decapitated female, (previously in Bed 53 (1914) near the Restaurant) is currently (1988) in the 'graveyard'. Having read Nan McDonald's poem, 'Her sheaf of wheat lay mildewed on her arm', and later David Campbell's 'A *Bather* struggling from her gown . . . gooseflesh pimpling stone' I am sure she is the statue they were referring to — one of the batch of Charles F. Summer's statues that came from Italy in 1883. The description in the *Sydney Morning Herald* article of 10 January 1883 — 'Summer, with her right hand encircling a sheaf of wheat and cornflower, while her left hand grasps a sickle', confirms that this is the same statue.

Late in the afternoon of 27 January 1989, about a week after Venus Italica was decapitated, I went in search of her 'Lost Summer' to find her mournful (and also decapitated) in the 'graveyard' behind the Succulent Garden.

The Death of Summer

The picket fence and the parade
of prelates, layabouts, and bards —
where dossers wrap themselves in words
and follow Summer to her grave,
and harvest grain in recompense
now Summer's honey breath is stayed,
the ghost of Gerrald's eloquence
should spike their heads upon the fence
denying Winter her requite —

and scalp prestige from indolence
who murder gods with sickle blade,
when languid days restrain the light —
and smell the honey all day long,
and rule the Garden with a song.[61]

Sundial

Source: From Grantham
Date/Place: 1927
Removed: —
Additional Information: Situated in the corner of the Palm Grove near the glasshouses — not to be confused with the sundial (1857) that stood at the point of intersection of the pathways east of the Choragic Monument (see p.88). This sundial consisted of a sheet of slate, without a gnomon, resting on a sandstone pillar. It came from Grantham, the old home of Admiral King, son of the third Governor of New South Wales, and was presented to the Gardens in 1927,[62] and eroded away so much that it was removed.

Terracotta Statues

Source: Italy
Date/Place: 1883
Removed: —
Additional Information: 'Mr Moore has also received some terracotta statues, medallions, and alto-relievos [sculptures in high relief], the first of which are suitable enough for display in the Gardens, while the others are eminently fit for exhibition in some public building.'[63] The terracotta statues referred to here may well be the one still on the top of the sandstone columns in Centennial Park.

Urns/Vases

(a) Artificial Stone
Source: Gift of Mr Pratten, Australian Marble Works
Date/Place: 1859 (Lower Garden)
Removed: 10 vases or urns in the Lower Garden in 1921
Additional Information: The artificial stone Grecian vases on wreathed pedestals and plinth, made from powdered sandstone and Portland cement, were the first of their kind manufactured in Sydney.[64]

Steps leading to site of current Rose Garden with Statues and urns, 1899, *RBG Library*.

(b) Cliveden
Source: Haddonstone Pty Ltd
Date/Place: Steps to Rose Garden 1987
Removed: 1989
Additional Information: Two of the four artificial stone urns on the Rose Garden Steps were vandalised on the evening of 1 November 1989.

(c) Marble
Source: —
Date/Place: 1903, Palace Garden/Palace Garden Steps
Removed: *c.* 1965
Additional Information: Several of these are in storage in 'graveyard' (1988), two having recently been placed on Palace Garden Steps.

There were bases for vases on side copings of the Palace Garden Steps, and at the four corners figures representing the Seasons.[65]

In 1909 the pedestals of the large marble vases at the side of the stone steps and on the western side of the circular rose bed (on the same walk) were reported as being out of perpendicular.[66]

Prior to the introduction of the Haddonstone Urns in 1987, the last urns in the Gardens (in bed opposite Bed 82) had been removed in 1965.

(d) Terracotta Urns
Source: Selected by Dr George Bennett, Doulton and Wall's yards in London

Date/Place: 1873/1879. Two terracotta urns once stood at entrance to Palm Grove
Removed: —
Additional Information: The Terracotta Vases of 1873, selected by Dr George Bennett from Doulton and Wall in 1873, were a later addition, following the gift of Mr Pratten's stone vases in 1859.[67] In 1879 Dr Bennett selected some (additional?) terracotta vases. One of them was located adjacent to the main gate of the Macquarie Wall in the Lower Garden. One of the (earlier?) Doulton terracotta vases (now painted) is currently on the entrance steps to the Visitor Centre.

Venus de Medici

Source: Art Gallery of New South Wales
Date/Place: 1903, Palace Garden between steps and Palm Grove — on left-hand side of pathway facing Palm Grove
Removed: Probably removed in the 'morality' purge of 1915 and placed in storage
Additional Information: 60-inch (approximately) marble after the antique. Mr Kirkpatrick, visiting the Gardens, had seen one of the employees destroying the statue of Apollo. He persuaded Mr Ward, the Superintendent, to allow him to take the statue of Venus de Medici, and he placed it in his garden at Lapstone. Subsequently it was placed in the grounds of the Lapstone Hotel where the statue was 'an object of admiration by thousands of visitors'.[68] In a rather scurrilous piece in *Smith's Weekly* the existence of the statue in the Gardens had apparently been denied by Mr Ward.[69]

The situation was complicated by a compensation claim by Mr John Emmans against the Gardens. The statue had apparently been moved into a packing-shed attached to the Gardens, along with three other statues. John Emmans said he had helped load the statue on a truck and 'put his old felt hat on her head and kissed her an affectionate good-bye'. It is hard to imagine such a quotation to be other than true.

The area of the Lapstone Hotel is now controlled by the Air Force and the fate of the statue is unknown.

Queen Victoria (see Albert the Good)

Source: Statue by sculptor Marshall Wood
Date/Place: 1879
Removed: Destroyed by fire in 1882
Additional Information: Statue of young Queen Victoria destroyed by burning of Garden Palace in 1882. The foundation stone was removed in 1897 to make the floor of the Rose Garden (Chess) Pavilion.

Whale's Head

Source: Washed up on Bondi beach some years prior to 1846
Date/Place: 1921, near Glasshouse No. 10
Removed: —
Additional Information: The 1921 map shows the Whale's Head near Glasshouse No. 10. The rockery around the Port Jackson Fig was at one time called the Whale's Head Rockery. 'Rocks and stones of every imaginable shape, searched for upon the coast in the neighbourhood, for the purpose — a portion of the head (skull) and a number of the vertebrae of a whale cast ashore some time ago at Bondi are seen, and from them sprouting almost every variety of cactus known in the colony'.[70]

Winter

Source: Job Hanson, 1879/Italy, 1883
Date/Place: Lawn 16, near Bed 100 in 1965
Removed: —
Additional Information: There were two statues of Winter, a small one and a large one. A large statue of Winter (an old bearded man with a dead bird is currently (1988) in storage). The other statue was transferred to the garden of the Chief Justice (maintained by the Gardens).[71]

(The) Wrestlers

Source: By I. Lottatori, donated by Art Gallery of New South Wales
Date/Place: Lower Garden, gift to Gardens 1958. Bed 51 in 1965
Removed: There in 1965. No trace, 1972. Destroyed by vandals?
Additional Information: Massive green marble (greenstone) sculpture from the Art Gallery (1501) 40 × 45 inches — formerly near Gardens Restaurant — original at Florence. Found AD 1535 near the Lantern, Rome. Original attributed to Cephissodotus, of the School of Rhodes, Fifth Period of Greek Art.[72]

8

Centennial Park

The Picket Fence and the Parade

Centennial Park was under the administration of the Royal Botanic Gardens from 1888 until 1980. From 1990 it again received administrative back-up from the Gardens in finance and personnel.

Henry Parkes commissioned Charles Moore to set up a Centennial Park to mark the centenary of European settlement of Australia.[1] 'Henry Parkes was responsible for securing [it] against the builders, and, typically, he envisaged a . . . mausoleum there, where the great of New South Wales (himself among them) should be entombed among the carriageways and artificial lakes'.[2]

According to Mr A. J. Clark, the citizens of Sydney raised a considerable amount of money in those early days to purchase six Italian marble statues. Two of these were The Boxers; the others being the seasons Spring, Summer, Autumn and Winter.[3] It is believed that the bulk of the remaining statues were 'standard ornamental pieces purchased from a catalogue' or obtained by an Australian delegation to an American Trade Fair in the 1890s.[4] This is supported by the fact that the bulk of the statues fall into two main categories:

1. Plaster on sandstone bases and terracotta statues (15) like the two (terracotta statues) on the Doric columns near the lake (inscribed 'Villeroy & Boch' — the famous Luxembourg pottery firm); and

2. 'Larger than life' (110%) marble statues (11), several of which have an American theme — Abraham Lincoln (President 1861–65); 'One More Shot' (a copy of a statue by the American sculptor, John Roger, who specialised in scenes from the Civil War 1861–65); and James Abram Garfield (President March-December 1881). Both Lincoln and Garfield were assassinated.

The smaller marble statues were probably imported by Mr Job Hanson (the same J. Hanson who won a medal at the International Exhibition), who was paid £1,970 0*s*. 0*d*. in 1889 for erecting 11 statues in Centennial Park).[5]

The famous bronze statue 'We Won' (better known as 'The Footballer', see p.142) was commissioned by Sir Henry Parkes in 1891. The sculptor was the Italian Tommaso Sani and it is his most important work.

The other important statue in the park was that of Sir Henry himself, significant more for its 'memorial qualities' than as a piece of art. It was the last sculpture to be placed in the park, in 1897, and remained in the same location until 1971, when it was destroyed by vandals.[6]

In the 1920s there were 31 statues in Centennial Park. Most of these had been destroyed by 1974.[7]

The first recorded act of vandalism on statues in Centennial Park occurred in 1898 when one of the Seasons was pushed over and badly broken, but skilfully repaired and replaced on its pedestal.[8] Considerable vandalism then followed in 1899 with the breaking off of arms, fingers, and other pieces of terracotta statues, apparently from stones being thrown at them.[9] By 1946 only 19 statues had survived.[10] In 1947 it was recommended that several of them be moved to new sites where they were less likely to be damaged by the increasing traffic. Most of the statues were finally removed in the early 1970s.

In 1988 only three were left — the renovated statue of 'The Footballer', and the two terracotta statues on the sandstone columns in the Rose Garden.

In the early 1930s there were large numbers of unemployed men living in the Park and many statues, including the two griffin statues (winged lions) at the entrance were damaged.[11] The vandalism was probably still caused by young males, and not necessarily the older (unemployed) men. A. J. Clark, Principal of Hurstville Boys High School in 1977, said he could remember being 'most impressed by (the) giant white statues when he was six'.[12]

Twelve statues were removed in 1971, and replacement

Centennial Park Panorama, 1892, *RBG Library*.

statues from the Museum(?) 'were all destroyed ... while the staff was having lunch'.[13] In December 1971 the marble sculpture of Sir Henry Parkes, after vandals had 'removed some of the ''Grand old Man's'' fingers among the vulnerable parts'[14] suffered the final ignominy of being completely destroyed by two bomb explosions, believed at the time to be the work of an extremist group.[15]

By January 1972 there were only 'eight broken white figures left ... of [the] statues that once dotted Centennial Park ... stored behind a wire fence of a workman's compound'.[16] Left behind were 'the battered remains of Gladstone, Dickens, Diana, the Greek Goddess of Hunting, and a scarred statue called The Boxer'.[17] The Park Superintendent, Mr Ron Salkeld, was quoted as saying that 'general larrikinism [has] increased ... over the past few years', and that he could 'see no way out but to have no more statues [as] trees and flowers grow back again and road signs can be re-erected ... but statues break to pieces'.[18] According to the *Daily Telegraph*: 'The only two statues likely to be restored were the Griffins ... the rest [were] to be kept in their [barbed wire] compound, known as ''the tip'' to Park employees'.[19]

About this time there were numerous requests regarding the purchase of broken statues from Centennial Park. The Sydney Branch of the Dickens Fellowship inquired if they could procure their patron, to which the Director, Dr Beard, replied: 'During the last few months we have moved all the statuary in the Park ... If your Society would like to obtain this [slightly damaged] statue for erection elsewhere, I shall be grateful for a written application [to] the Minister for Agriculture.'[20]

One of the departed, William Gladstone (1809–98) is reported to have once said (of the departed): 'Show me the manner in which a nation cares for its dead and I will measure with mathematical exactitude the tender sympathies of its people, their respect for the laws of the land, and their loyalty to high ideals'.[21] The same could well be said of statues!

Where once ... 31 statues had stood there were now only two remaining 'perched on tall columns beyond the reach of vandals'.[22] One of these was a 'female figure whose identity the park authorities seem to have forgotten'.[23] The other was 'a philosopher or senator [*sic*] ... [on top of a] tall stone pillar ... in the Egyptian style'.[24]

Both the statues on the columns are females and it is not known what they represent. They both have the name Villeroy & Boch inscribed on them and were installed in the 1890s. It is believed that the fluted Doric columns on which they stand came from the Darlinghurst Court House, although the Court House columns, of what appears to be the same red sandstone, do not have Acanthus capitals like those in Centennial Park.[25] These could easily have been carved separately at the time. The two columns and their statues, far removed from the stones of little boys and other temptations, were still there in 1988.

Alphabetical Checklist of Statues and Memorials from Centennial Park

Architecture

Source: Possibly Villeroy & Boch
Date/Place: 1890s, Carrington Drive
Removed: Removed prior to 1946
Additional Information: Plaster on Sandstone Base.

Autumn

Source: Italy(?) Probably imported by Job Hanson
Date/Place: 1890s, Parkes Drive at Rose Garden
Removed: Removed *c.* 1970
Additional Information: Marble. Remaining in 1946, paint peeling off.

The Boxers

Source: Italy(?) Probably imported by Job Hanson
Date/Place: 1890s, Queen Street Gate
Removed: One scarred statue left in 1972, one went to Hurstville Boys High School in 1976
Additional Information: Marble on stone base.

Busby's Bore Cairn

Source: Constructed in 1988
Date/Place: 1988, inside Cleveland Street Gate
Removed: —
Additional Information: To commemorate the location of Busby's Bore.

Cannon

Source: Russian Cannon, siege of Sebastopol, Crimean War, 1854–55
Date/Place: Near We Won statue, from *c.* 1920, from Botanic Gardens
Removed: Still there 1988
Additional Information: Two cannon mounted on wheels with Russian inscriptions. An inspection of an old photograph of the Bourke statue has confirmed that these are the same guns (see p.109).

The symmetrical inscriptions (in Slovanian Russian script) read as follows:

	Cannon one	Cannon two
Side A	36 phunt [weight measure] [illegible]	weight 77 [or 177] pounds, 98 phunts January 1852
Side B	No. 31,585 Alexander's Factory Supervisor L. [?] Butinev	No. 31,523 Alexander's Factory Supervisor L. [?] Butinev
top/back of cannon	M.A. Moscow Artillery [?]	M.A.

These two Russian cannon that would have killed and maimed so many men look almost beautiful, like works of sculpture, so incongruously flanking an Italian footballer surrounded by eight lions.[26]

Charity

Source: Possibly Villeroy & Boch
Date/Place: 1890s, Grand Drive near Kiosk
Removed: Removed prior to 1946
Additional Information: Plaster on sandstone base. Considered to have 'more sophisticated modelling'.[27]

Sebastopol Cannon, 1992, *Jaime Plaza.*

(Statues on) Columns

Source: Statues have name Villeroy & Boch on them
Date/Place: Near lakes, placed there 1890
Removed: Still there 1988 — only statues to have remained permanently on site
Additional Information: Terracotta statues of two women on fluted Doric sandstone columns (with *Acanthus* capitals) from old site of Australian Museum (Darlinghurst Court House?).

Commerce

Source: Possibly Villeroy & Boch
Date/Place: 1890s, on Carrington Drive
Removed: Removed prior to 1946
Additional Information: Plaster on sandstone base.

Diana

Source: Possibly Job Hanson
Date/Place: 1890s, Queen Street Gate
Removed: Extant in 1946 with a few chips off statue. Removed *c.* 1970
Additional Information: Marble on sandstone base.

Diana the Huntress

Source: Possibly Job Hanson
Date/Place: Grand Drive and branch road leading to Queen Street
Removed: Badly damaged *c.* 1958. Removed 1972
Additional Information: Marble. Set back 40 feet in 1911.[28]

Dickens

Source: Unknown
Date/Place: There prior to 1897. On Dickens Road
Removed: 1972. In storage
Additional Information: Marble. Extant in 1946 with broken nose and several chips. Marble statue of Charles Dickens. Moved in 1897, when the statue of Sir Henry Parkes was erected, to ... 'the junction of what are known as the flat-roads ... with little statuary within a considerable distance'.[29]

Charles Dickens, *c.* 1890, *RBG Library*.

Disraeli

Source: Unknown
Date/Place: 1890s, on Grand Drive near Cleveland Street exit
Removed: Destroyed by accident pre-1946
Additional Information: Marble on sandstone base of English politician, Benjamin Disraeli.

Federation Monument

Source: Commonwealth Stone created in 1904
Date/Place: Site of actual ceremony of swearing in of the first Governor-General of Australia
Removed: Still there today. Federation Pavilion constructed in 1988
Additional Information: The ceremony associated with the Proclamation of the Federation of Australia took place in Centennial Park on 1 January 1901. It was reported as being a scene 'of inspiring grandeur with 60,000 attending'.[30] A temporary plaster of Paris pavilion

was constructed for the ceremony. This octagonal, domed building, 14 metres high — richly decorated with bas-relief castings of native flora and the imperial coat of arms — was removed in 1903.[31]

The Commonwealth Stone (or Monument to Federation) was created in 1904 when 'the stone on which the first Governor-General [of Australia — the Earl of Hopetoun] was sworn in, was erected on a suitable pedestal on the site of the actual ceremony.[32] The pedestal was then surrounded by a stone coping, surmounted by an iron railing. The stone, a massive polished block of Moruya granite, in the form of a hexagon, had its sides inscribed to represent each of the federating states.[33]

The dates inscribed on the monument are as follows:

N.S.W., 1788 — First permanent settlement
Victoria, 1851 — Year of formation into separate colony
Queensland, 1859 — Year of formation into separate colony
South Australia, 1836 — First permanent settlement
Tasmania, 1825 — Year of formation into separate colony
Western Australia, 1825.

There is a problem with the date for Western Australia. Efforts had been made to establish a penal colony there, under Major Lockyer, in 1826.[34] The first attempt at permanent settlement was not until 1829.

A permanent Federation Pavilion, designed by Alexander Tzannes, was constructed on the site in 1988. Its dome creates a 'lantern' effect, with light being introduced from both the perimeter and an oculus. The interior of the dome was embellished by Imants Tillers. Around the outside of the sandstone frieze below the dome, the words 'Mammon or Millennial Eden' are inscribed. They come from the sonnet, 'Australia', written by Bernard O'Dowd in 1901.

Australia

Last sea-thing dredged by sailor Time from Space,
Are you a drift Sargasso, where the West
In halcyon calm rebuilds her fatal nest?
Or Delos of a coming Sun-God's race?
Are you for Light, and trimmed, with oil in place,
Or but a Will o' Wisp on marshy quest?
A new demesne for Mannon to infest?
Or lurks millennial Eden 'neath your face?[35]

The landscaping of the site incorporates an elevated clearing from which the pavilion may be viewed to create a scale reflecting the rural areas of Australia, while allowing the space to be used on an everyday basis, and as a setting for major ceremonial events.[36]

Fine Arts

Source: Probably imported by Job Hanson
Date/Place: 1889, along High Road
Removed: Removed prior to 1946
Additional Information: Marble.

The Footballer (see We Won)

Garfield

Source: Unknown
Date/Place: 1890s, Grand Drive — at corner of road leading to York Street
Removed: Extensively damaged by 1946 and removed *c.* 1958
Additional Information: Marble, currently in compound. James Abram Garfield was President of the United States of America from March 1881. Shot in the face in September of that year, he died in December.

The Arts, *c.* 1890, *RBG Library*.

Gladstone

Source: Unknown
Date/Place: There before 1897, at Cleveland Street Gates near Cook Road
Removed: 1946 nose and toe broken off, several chips off fingers. Extensively damaged *c.* 1958. Removed 1974
Additional Information: Marble statue. 1989, Restored T. A. Taylor's and transferred to Gladstone city council, Queensland.

Griffins

Source: Possibly Villeroy and Boch
Date/Place: 1890s, on main entry road from Paddington on either side of Ceremonial Triangle
Removed: Damaged in 1930s and 1946 and put into storage
Additional Information: Plaster(?) on stone base. One of the Griffins lost a wing in 1900, 'possibly through a defect in the stone' (*sic*).[37]

Industry

Source: Possibly Villeroy & Boch
Date/Place: 1890s, at Randwick Gates
Removed: 1970(?) Remaining in 1946, thumb broken, paint peeling off
Additional Information: Plaster on sandstone base.

Hamilton (Brolga) Memorial Gates

Source: Installed by the Gould League
Date/Place: *c.* 1937
Removed: Still in place
Additional Information: Memorial to Harold Hamilton, first secretary of the Gould League. The adjacent bronze figures were designed by Neville Cayley. They were originally near the paper-bark swamp near the lily pond.[38]

Lincoln

Source: Unknown
Date/Place: 1890s, Grand Drive — at corner of road leading to Ocean Street
Removed: Extensively damaged by 1946 and removed *c.* 1958
Additional Information: Marble on sandstone base of Abraham Lincoln, President of the United States of America, 1861–65 (assassinated).

Marble Lions

Source: Italy
Date/Place: Palace Garden Grounds Royal Botanic Gardens. Moved to Centennial Park during construction of Cahill Expressway
Removed: Centennial Park compound 1972
Additional Information: Two marble lions on sandstone bases. It is the intention of Professor Chambers to restore these two lions and return them to the region of the Palace Garden.

One More Shot

Source: Copy of statue by John Roger, an American sculptor who specialised in Civil War scenes
Date/Place: 1890s, corner of road leading to Market Street
Removed: Extensively damaged in 1946 and removed *c.* 1958
Additional Information: Marble on sandstone base. Standing figure with rifle and kneeling figure.

Painting

Source: Probably imported by Job Hanson
Date/Place: 1889, on Grand Drive
Removed: Destroyed by accident, possibly pre-1917.
Additional Information: Marble.

Parkes

Source: Erected August 1897
Date/Place: 150 metres from Oxford Street entrance to Park at Queen Street end of Memorial Triangle
Removed: Blown up by vandals, December 1971
Additional Information: Memorial to Sir Henry Parkes. 'The statue is a very good likeness . . . as he was about fifteen or twenty years before his death.'[39]

One More Shot, *c.* 1890, *State Archives Office.*

Queen Street Gate Plaque

Source: Australian Gas Light Company
Date/Place: 7 September 1962, Queen Street Gate
Removed: —
Additional Information: The lights on these gates were restored to commemorate the 125th anniversary of the Australian Gas Light Company.

Rotary Memorial Chair

Source: Rotary Club of Sydney
Date/Place: 26 September 1950
Removed: —
Additional Information: Unveiled by Angus Mitchell (Past President of Rotary International) to commemorate the planting of two fire-wheel trees — the eastern one in memory of Rotary founder Paul Harris, the western one in appreciation of the work of Angus Mitchell during his term as president.

Science

Source: Possibly Villeroy & Boch
Date/Place: 1890s, on Grand Drive, near old Tram Depot
Removed: Removed prior to 1946
Additional Information: Plaster on sandstone base.

Sculpture

Source: Possibly Villeroy & Boch
Date/Place: 1890s, on Carrington Drive — near Ocean Street Gate
Removed: Removed prior to 1946
Additional Information: Plaster on sandstone base.

Spring

Source: Italy(?) probably imported by Job Hanson
Date/Place: 1889, Parkes Drive at Rose Garden
Removed: Removed *c.* 1970
Additional Information: Marble. There in 1946, arm broken, body cracked, paint peeling off.

Spring, *c.* 1890, *RBG Library.*

Summer, *c.* 1890, *RBG Library.*

We Won, 1992, *Jaime Plaza.*

Summer

Source: Italy(?) probably imported by Job Hanson
Date/Place: 1889, in Rose Garden
Removed: Removed *c.* 1970
Additional Information: Marble. Still there in 1946, arm, broken, paint peeling off.

Urns/Vases

Source: Rofde & Andre (Importers)
Date/Place: 1895 (purchased in 1889), near Memorial Triangle
Removed: Removed by 1917
Additional Information: Terracotta vases on sandstone base — selected by Colonial Secretary in 1889. 'New Greek' with pedestals, 65. Pyrmont sandstone bases, £6 7*s.* 0*d.* One pair vases, no pedestals, 40. Pyrmont sandstone bases, £15 6*s.* 0*d.*

We Won

Source: Commissioned by Sir Henry Parkes of Italian sculptor Tommaso Sani (Florence), working in Australia in the 1880s
Date/Place: Erected 1893 near Oxford Street Entrance
Removed: Arm broken and in storage in 1972. Restored by Fine Art Bronze foundry, funded by Eastern Suburbs Junior Rugby League and the Centennial Park Trust, 9 April 1988
Additional Information: Life-sized bronze sculpture of young man on a pedestal originally surrounded by eight bronze lions on plinths, linked together by a brass chain. Believed to be the only statue of a footballer. (Rugby, not soccer) in the world.

Sani was working in Australia in the 1880s on the embellishment of public buildings — the best known of his works being the Pitt Street façade of the Sydney Post Office. His work aroused intense antagonism at the time. One panel showed a postman handing a letter to a smiling young woman, but: 'Critics maintained the woman was not smiling at the postman, but positively ogling him'. They also said a sheep was being sheared from the wrong end! State parliament appointed an expert committee. The sculptures were pronounced as unsuitable for display on public buildings and only saved by a spirited defence by the architect James Barnet.[40]

We Won is a wonderfully satirical piece of work with a 'life-sized figure dressed in a woollen cap, sweater, and tight knee-length trousers worn by the football players of the day. The face of the figure . . . [is very 'Italian' and] serene, and [an] idealised Neoclassical Apollo.'[41] In a letter to Sir Henry Parkes requesting payment on behalf of the artist James Barnet, Colonial Architect, describes the statue as follows:

> The figure (which is six feet in height) is nearly complete in plaster . . . and pleasing in its expression of Victory. The Pedestal with its four little girls representing respectively, Glory with Victor's wreath, Fame sounding its Trumpet, History, reading the deeds, and Modesty, looking on with approval, are all cleverly posed and expressive. Between these four Babies are Basso Relievo, representing the game of Football in four stages, such as 'Throwing', 'Scrimmage', 'Free-Kick' and 'Hurrah' — each one full of life and expression, in fact the whole when complete will be like a dream in Bronze.[42]

And indeed it is.

This is a valuable description, as Barnet would have no doubt discussed the symbolism of the work with the artist.

The statue, mounted on a pedestal, was originally surrounded by eight bronze lions on plinths, linked together by a brass chain.

In 1946 the football went missing and two lions were loose on their pedestals. The brass chain, then valued at £49, was stolen *c.* 1955. Following a letter from Mr Cedrick Flower it was reported that 'two of [the] lions have since been stolen. Bronze lions of an almost identical appearance . . . were sighted in a nearby antique dealer's shop. The matter was subsequently investigated by police . . . but these statues were replicas.'[43]

Meanwhile an arm had been broken from the statue and it was placed in storage *c.* 1970, to be restored and returned to its original position in 1988. The missing lions were also replaced but it was decided not to return the brass chain, as it would be too easily stolen.

Winter

Source: Italy(?), probably imported by Job Hanson
Date/Place: 1889, Parkes Drive at Rose Garden
Removed: Only pedestal remaining in 1946
Additional Information: Marble.

Federation Celebrations in Farm Cove prior to procession to Centennial Park, *Government Printing Office.*

Notes

Introduction

[1] Patrick White, *Voss*, Penguin, Harmondsworth 1963, p. 440.
[2] Lionel Gilbert, *The Royal Botanic Gardens Sydney: A History 1816-1985*, Oxford University Press, Melbourne 1986, p.67.
[3] RBG File 7(A).
[4] The poem, which was inspired by the Colossi of Egypt, was published by Leigh Hunt in *The Examiner*, January 1818, and reprinted with Rosalind and Helen in 1819.
[5] David Churches, *The Cultural Significance of the Royal Botanic Gardens, Sydney*, 1990, p. 158.
[6] *Sydney Morning Herald*, 27 September 1870.
[7] Rupert Gunnis, *Dictionary of British Sculptors, 1660-1851*, Abbey Library, London, 1968, p. 378.
[8] Churches, *op. cit.*, 10.1.
[9] Avenel Mitchell *et al.*, Art Gallery of New South Wales, Australian Sculpture, 1890-1919, Project Exhibition, Court 9, 22 January-22 February 1987.
[10] Joanna Mendelssohn, 'The Fashion of Fantasy', *The Bulletin*, 31 October 1989.
[11] Research by Judith Hunter: *Sydney Mail*, 27 September 1879; *Town and Country Journal*, 27 September 1879.
[12] 193-194, 196-197 Parkes Correspondence, Vol. 47, CYA917.
[13] *Ibid.*
[14] *Sydney Morning Herald*, 'News of the Day', 10 January 1883. Photocopy newspaper cutting 'discovered' by Bill Chapman and given to me by Doug Benson. The date and source were finally tracked down by Tony Jeans from other events reported on the page.
[15] From another lead by Doug Benson — Exhibition notes from Elizabeth Callaghan, Collections Manager Rotorua Art and History Museum (formerly The Bath-House). The 13 marble sculptures in their collection, in Carrara marble — the material used by Michaelangelo — purchased in 1908 for ¢2,000, are the largest collection of the works of Charles Francis Summers in the world. His statues may also be found in the Public Gardens in Ballarat and the gardens of the Exhibition Building, Melbourne.
[16] Ken Scarlett, *Australian Sculpture*, Nelson, West Melbourne, 1980, p. 632; from Sir Henry Parkes correspondence, letter 141, Vol. 39, Q709.901/M.
[17] RBG File 7(A), letter to Officer-in-Charge Botanical [*sic*] Gardens, from Miss T. Gidley, Principal, Kelvin Grove 'Private School', Oyster Bay, 19 August 1964.
[18] Churches, *op. cit.*, p. 791; J. H. Maiden, evidence to the Royal Commission on the improvement of Sydney, 1909.
[19] Extract from *Hansard*, 20 September 1966.
[20] RBG File 7(A), letter of reply by K. Mair, Director and Chief Botanist, to Mrs P. Wallis of Vaucluse.
[21] Note pinned to Solander bust in Gardens, handed on to me by Gary Chapple.
[22] J. H. Maiden, Evidence to the Royal Commission on the improvement of Sydney, 1909.
[23] May Gibbs, *The Complete Adventures of Snugglepot and Cuddlepie*, Angus & Robertson, Sydney, 1963, p. 2.
[24] *Sun*, 23 May 1930. Researched by Churches, *op. cit.*, p. 790.
[25] *Sydney Morning Herald*, 30 September 1978.
[26] RBG File 7(A).
[27] RBG File 7(A), letter from Dr L. A. S. Johnson to Mr J. Morris, Assistant Director, The National Trust of Australia.

1 The Governor's Domain

[1] John Hunter, *An Historical Journal of Events at Sydney and Sea, 1787-1792*, RAHS reprint, Sydney 1968, pp. 41, 42.
[2] David Collins, An Account of the English Colony in New South Wales, T. Cadell & M. Collins, London 1804, pp. 468, 470, 474.
[3] Geeves on Thursday, *Sydney Morning Herald*, 16 October 1980.
[4] Watkin Tench, *A Complete Account of the Settlement at Port Jackson in New South Wales*, RAHS reprint, Sydney 1961, p. 156.
[5] Roy H. Goddard, *The Life and Times of James Milson*, Georgian House, Melbourne 1955, p. 35 — description of farmers' huts in Hawkesbury in 1910.
[6] J. F. Campbell 'Historical Notes on Government House Domain, Sydney', *JRAHS*, Vol. XVII, Part II, 1931, pp. 101-126 — Mandate dated 2 December 1792.
[7] 'Plan of the Town of Sydney in N.S.W. by James Meehan', *Historical Records of N.S.W.*, Vol. VI, p. 366.
[8] R. H. Cambage (ed.) from Notes of the late J. H. Maiden, 'History of the Sydney Botanic Gardens', *JRAHS*, Vol. XIV, Part I, 1928, p. 26.
[9] Governor Hunter's Dispatch, 30 April 1796.
[10] Lionel Gilbert, *The Royal Botanic Gardens, Sydney: A History 1816-1985*, Oxford University Press, Melbourne 1986, p. 15.
[11] Research by Mr Ted Higginbotham for the *Sydney Harbour Tunnel Newsletter*, 1989, p. 3.

[12] *The Sydney Gazette and New South Wales Advertiser*, Sunday 29 September 1805, No. 135, p. 2 (b–c), Angus & Robertson and Trustees of the Public Library of N.S.W., Sydney 1966.
[13] *Ibid.*, Sunday 17 November 1805, No. 142, 2(c); Sunday 29 December 1805, No. 148, 2(b).
[14] *Sydney Gazette*, 29 May 1808.
[15] J. F. Campbell, *JRAHS*, Vol. XVII, Part II, 1931, pp. 115–18.
[16] Historical Feature, 'Botanic Gardens' Troubled Origins', *Daily Telegraph Mirror*, 19 October 1990.
[17] *Ibid.*
[18] David Churches, *The Cultural Significance of the Royal Botanic Gardens, Sydney*, 1990, p. 21.
[19] *Daily Telegraph Mirror*, 19 October 1990.
[20] Susanna de Vries-Evans, *Historic Sydney as Seen by its Early Artists*, Doubleday, Sydney 1983, map derived by Bryan and Zillah Thomas published on endpapers.
[21] Philip Geeves, *Philip Geeves' Sydney*, Angus & Robertson, Sydney 1981, p. 56.
[22] D. J. Mulvaney, *A Good Foundation — Reflections on the Heritage of the First Government House, Sydney*, Australian Government Publishing Service, Canberra 1985.
[23] Cambage, *op. cit.*, p. 28.
[24] *Ibid.*, from *HRNSW.*, Vol. VI, from Jas. Meehan's map of 31 October 1807 attached to p. 367.
[25] *HRNSW*, Vol. VI., map attached to p. 367.
[26] *Ibid.*, p. 347.
[27] *Daily Telegraph Mirror*, 19 October 1990.
[28] Cambage, *op. cit.*, p. 29.
[29] *Ibid.*, pp. 29–30.
[30] Gilbert, *Royal Botanic Gardens, Sydney*, p. 20; *The Voyage of Governor Phillip to Botany Bay...*, London 1789, p. 48.
[31] *Daily Telegraph Mirror*, 19 October 1990.
[32] *Ibid.*
[33] L. A. Gilbert, *Botanical Investigation of New South Wales, 1811-1880*, PhD thesis, University of New England, 1971, p. 440.
[34] *Daily Telegraph Mirror*, 19 October 1990.
[35] Cambage, *op. cit.*, pp. 30, 32.
[36] *Ibid.*, pp. 30–31, from earlier research of the late Mr Critchett Walker in the records of the Colonial Secretary's office.
[37] *Daily Telegraph Mirror*, 19 October 1990.
[38] Cambage, *op. cit.*, p. 32, from *HRNSW*, Vol. VII, p. 634.
[39] *Ibid*, p. 33; Campbell, *op. cit.*, p. 122.
[40] Cambage, *op. cit.*, p. 36.
[41] *Daily Telegraph Mirror*, 19 October 1990.
[42] Cambage, *op. cit.*, p. 31. The Government Domain was enclosed with a stone wall, dividing it from the town across the neck of land between 'Sydney Cove' and Woolloomooloo Bay — work No. 46 of the public works undertaken by Macquarie, see his report dated 27 July 1822, to Earl Bathurst.
[43] Research by Shirley Colless, *HRA*, 15 December 1817, Series I, Vol. IV, p. 375.
[44] *Daily Telegraph Mirror*, 19 October 1990.
[45] Cambage, *op. cit.*, pp. 33-34.
[46] J. H. Maiden 'Sydney Botanic Gardens', *Royal Botanic Gardens Kew Bulletin*, London, No. 6, 1906, p. 216.
[47] Leigh & Co's *Handbook to Sydney and Suburbs*, 1867, p. 24.
[48] Maiden, 'Sydney Botanic Garden', Royal Botanic Gardens Kew Bulletin, *op. cit.*, p. 37.
[49] *Sun*, 21 January 1914.
[50] *Ibid.*
[51] J. P. McGuanne, 'Bennilong [*sic*] Point and Fort Macquarie', *JRAHS*, Vol. I, Part 2, 1901, pp. 9–13.
[52] Leigh & Co., *op. cit.*, p. 24.
[53] Churches, *op. cit.*, p. 157; *HRA*, Vol. X, p. 687.
[54] Churches, *op. cit.*, draft manuscript, 14.03; A. Cunningham to R. W. Hay, 10 July 1832, AO.
[55] Walter W. Froggatt (ed.) from Notes of the late J. H. Maiden, *JRAHS*, Vol. XVII, Part III, 1931, p. 153.
[56] *Ibid.*
[57] *Ibid.*
[58] Churches, *op. cit.*; Froggatt, *op. cit.*, p. 153.
[59] J. F. Campbell, 'Notes on the Historical Development of Macquarie Street South, 1810 to 1880', *JRAHS*, Vol. 23, Part 3, 1937, p. 200.
[60] *Sun*, 21 January 1914.
[61] Indenture of release from Mortgate between Mortimer William Lewis and George John Rogers, 17 January 1850, Society of Australian Genealogists, S.A., 342; *Sand's Sydney Directory for the Year 1850.*
[62] Letter from L. A. Punch, to K. A. Johnson, President of the Society of Australian Genealogists, 12 December 1978, re Country Party occupation of Richmond Villa.
[63] *Ibid.*
[64] BG/AR, 1849, p. 2.
[65] BG/AR, 1871, p. 4; Charles Moore to Legislative Assembly, 1870-71, State Archives, Supplement No. 9.
[66] Froggatt, *op. cit.*, p. 154.
[67] BG/AR, 1878/79, p. 4.
[68] Churches, *op. cit.*, p. 156.
[69] *Ibid.*
[70] Froggatt, *op. cit.*, p. 128.
[71] *Ibid.*, p. 3.
[72] BG/AR, 1981/82, p. 30.
[73] Research by Robert Staas, architect, Noel Bell, Ridley Smith & Partners Pty. Ltd.
[74] J. H. Maiden, 'Sydney Botanic Gardens', photocopy of an article in my possession of unknown source, p. 217.
[75] *Ibid.*
[76] Mrs Charles Meredith, *Notes and Sketches of New South Wales*, Penguin Colonial Facsimiles, Ringwood 1973, p. 39.
[77] BG/AR, 1903, p. 21.
[78] Froggatt, *op. cit.*, p. 153.
[79] Churches, *op. cit.*, draft manuscript, 9.02; (Fraser to Goulburn, 14 January 1822, AO 4/1752, p. 26).
[80] *Sydney Morning Herald*, 26 June 1846.
[81] BG/AR, 1899, p. 3.
[82] BG/AR, 1855, p. 2.
[83] Froggatt, *op. cit.*, p. 153; Charles Moore, report to Legislative Assembly, 1870-71 (State Archives, Supplement No. 9).
[84] Gilbert, *Royal Botanic Gardens*, Sydney, p. 104; from *Sydney Mail*, 16 September 1871.
[88] BG/AR, 1899, p. 3.

[89] BG/AR, 1916, p. 5.
[90] BG/AR, 1910, p. 3.
[91] Churches, *op. cit.*, p. 97.
[92] Commonwealth Scientific and Industrial Research Organisation, 1949, a development of the Council for Scientific and Industrial Research, established 1926.
[93] Told to me by Peter Hind, seed collector, Royal Botanic Gardens Sydney.
[94] Patrick White, *Voss*, Penguin, Harmondsworth 1963, p. 46.
[95] Tom Carter, *The Victorian Garden*, Bell & Hyman, London 1984, p. 7.
[96] *Ibid.*, p. 109.
[97] BG/AR, 1906, p. 4.
[98] BG/AR, 1915, p. 5.
[99] *Sun*, 21 January 1914; *Illustrated Sydney News*, 21 November 1854.
[100] Froggatt, *op. cit.*, p. 154.
[101] *Sydney Morning Herald*, 10 January 1878. Information supplied by Mrs N. Sommers, a relative of Gardens' Director Charles Moore.
[102] BG/AR, 1898, p. 15.
[103] B. C. Peck, *Recollections of Sydney*, London 1850 pp. 60, 64–65; from Frank Crowley, *A Documentary History of Australia*, Vol. 2, *1841–1874*, Nelson, Melbourne 1980, pp. 156–158.
[104] J. Askew, *A Voyage to Australia and New Zealand*, London 1857, pp. 196–206; *Illustrated Sydney News*, 21 November 1854.
[105] BG/AR, 1902, p. 25.
[106] An Act for Regulating the Police in the Town . . . of Sydney, 6 August 1833, 4 GUL. IV No. vii.
[107] 'The Harbour and Bathing', *The Saturday Magazine*, 1 April 1837, p. 123.
[108] *Household Words*, 11 November 1850.
[109] Askew, *op. cit.*, pp. 196–206 (referring to a visit in July 1855).
[110] *Sun*, 21 January 1914.
[111] Leigh & Co, *op. cit.*, p. 24.
[112] BG/AR, 1898, p. 22.
[113] R. Brasch, *How Did Sport Begin?* (2nd edn), Collins Australia, Sydney 1989, p. 346.
[114] *Ibid.*
[115] *Ibid.*
[116] BG/AR, 1900, p. 21.
[117] *Australia's Yesterday's: A Look at our Recent Past*, Reader's Digest, Sydney 1974, p. 192.
[118] BG/AR, 1902, p. 26
[119] Information supplied by Bernard Connell, N.S.W. Public Works Department.
[120] BG/AR, 1904, p. 21; BG/AR, 1898, p. 24.
[121] BG/AR, 1908, p. 25.
[122] BG/AR, 1909, p. 30.
[123] Gilbert, *Royal Botanic Gardens*, Sydney, p. 143, from Rollo Gillespie, *Vice Regal Quarters*, Angus & Robertson, Sydney 1975, pp. 70–72; Annual Reports, Botanic Gardens, June 1924, and June 1925; Churches, *op. cit.*, 9.05; *HRA* Vol. XIX, Part I, pp. 44–47.
[124] BG/AR, 1905, p. 29.
[125] State Archives Supplement, No. 9.
[126] BG/AR, 1848, p. 1.
[127] Article published by Taronga Zoo to celebrate its seventy-fifth birthday.
[128] Froggatt, *op. cit.*, p. 161.
[129] Article published by Taronga Zoo to celebrate it seventy-fifth birthday.
[130] *Ibid.*
[131] *Ibid.*
[132] Froggatt, op. cit., p. 162; quotation from the *Sydney Mail*, 8 September 1860.
[133] Leigh & Co, *op. cit.*, p. 25.
[134] *Ibid.*
[135] *Ibid.*
[136] BG/AR, 1878/79, p. 2.
[137] BG/AR, 1916, p. 6.
[138] Told to me, 7 November 1990, by Mrs Daphne Robilliard. Her great grandfather James McFarland was Head Gardener from about 1875 and lived in Cottage 1 (Cottage 4?).
[139] Froggatt, *op. cit.*, p. 161.
[140] BG/AR, 1910, p. 4.
[141] Interview with Sid Morrison by Heather Edwards, 26 June 1990. Sid Morrison retired in 1958.
[142] *c.* 1985 I had a call from a Gwen Jacobson, then working for Bay Books, who said her great great grandfather, a Captain Venus (*sic*), had brought a tortoise from the Solomon Islands to the Gardens.
[143] I was informed by Harry Craig, grandson of J. H. Maiden, that this tortoise had once laid an egg.
[144] *Illustrated Sydney News*, 15 May 1880, pp. 4, 6.
[145] Edwin Wilson, 'The Garden Palace — Its Rise and Fall', *Gardenscene*, February 1987; costs and numbers of visitors provided by David Churches.
[146] Gilbert, *Botanical Investigation*, p. 520 (n).
[147] Walter W. Froggatt 'The Curators and Botanists of the Botanic Gardens', *JRAHS* Vol. 18, Part 3, 1933, pp. 101–133; Annual Report, Botanic Gardens, 1915, p. 37.
[148] BG/AR, 1915, p. 37.
[149] *The Australian Encyclopaedia* (4th edn), Grolier Society of Australia, Sydney 1983, Vol. 7, p. 125.
[150] *Illustrated Sydney News*, 12 July 1879, pp. 4, 10.
[151] Coloured map of International Exhibition of 1879, lithographed, New South Wales Surveyor-General's Office.
[152] Minute re Government Domain and Botanic Gardens to Chief Secretary's Office, Sydney, C.W., P.U.S., 30 January 1896, State Archives, Supplement No. 9.
[153] A. G. ('Johnnie') Moyes, *Australian Cricket; A History*, Angus & Robertson, Sydney 1959, pp. 104, 111. The next match and the four after that were won by Victoria.
[154] Edwin Wilson (ed.) Shirley Colless (research assistant), *Discovering the Domain*, Hale & Iremonger, Sydney 1986, p. 24; *Sydney Morning Herald*, 16 December 1936; *Sydney Morning Herald*, 30 January 1862 (*Sydney Morning Herald* History Supplement, 8 September 1981).
[155] *Sun*, 21 January 1914.
[156] BG/AR, 1908, p. 25.
[157] *Sun*, 21 January 1914.
[158] Information supplied by Alan Tillsley, Royal Botanic Gardens Sydney.

2 Roads and Gates

[1] BG/AR, 1918, p. 3.
[2] R. H. Cambage (ed.) from Notes of the late J. H. Maiden, 'History of the Sydney Botanic Gardens', *JRAHS*, Vol. XIV, Part I, 1928, p. 34.
[3] Lionel Gilbert, *The Royal Botanic Gardens, Sydney: A History 1816–1985,* Oxford University Press, Melbourne 1986, p. 25. The actual quotation comes from Macquarie's Diary, ML.A. 773, p. 25.
[4] J. H. Maiden, 'Sydney Botanic Gardens', *Royal Botanic Gardens Kew Bulletin*, London, No. 6, 1906, p. 212.
[5] *Sydney Morning Herald*, 26 June 1846.
[6] Cambage, *op. cit.*, p. 41.
[7] BG/AR, 1902, p. 27.
[8] Edwin Wilson (ed.), *Royal Botanic Gardens Sydney*, Royal Botanic Gardens and Domain Trust, Sydney 1982, p. 11.
[9] Edwin Wilson (ed.) Shirley Colless (research assistant), *Discovering the Domain*, Hale & Iremonger, Sydney 1986, p. 12.
[10] BG/AR 1899, p. 24; BG/AR, 1905, p. 26.
[11] Research by Robert Staas, Noel Bell, Ridley Smith & Partners Pty Ltd.
[12] Leigh & Co's *Handbook to Sydney and Suburbs*, 1867, p. 24.
[13] Wilson (ed.), *Discovering the Domain*, p. 24.
[14] BG/AR, 1912. From old photographs it can be established that the gates were there in 1842 when the Bourke statue was unveiled.
[15] Walter W. Froggatt, (ed.), from Notes of the late J. H. Maiden, Vol. XVII, Part II, 1931, p. 127.
[16] Charles Davis, 'A Stroll through the Botanical Gardens', *Sydney Quarterly Magazine*, September 1888, p. 210.
[17] Froggatt, *op. cit.*, pp. 127–8.
[18] *Ibid.*, p. 128; Gilbert, *op. cit.*, p. 104; David Churches, *The Cultural Significance of the Royal Botanic Gardens, Sydney*, 1990, p. 722, quotes the cost as being ¢733 4*s*. 0*d*.
[19] RBG File 7(A), Minute K. Mair to Director-General of Agriculture, 9 December 1965.
[20] Froggatt, *op. cit.*, pp. 133–134.
[21] Churches, *op. cit.*, p. 722.
[22] BG/AR, 1912, p. 40; Churches, *op. cit.*, 11.3.2.
[23] J. H. Maiden, *Guide to the Botanic Gardens*, Government Printer, Sydney 1903, p. 2.
[24] BG/AR, 1967 (as part of the Annual Report for the N.S.W. Department of Agriculture) p. 119.
[25] Davis, *op. cit.*, p. 210.
[26] RBG File 7(A). Letter from Jill Hellyer to the Minister for Agriculture.
[27] Information supplied by Warwick Watson. Churches, *op. cit.*, p. 113, indicates this gate was originally called the 'Governor's Private Gate' because it gave access to the Governor's grounds prior to the rearrangement of the boundaries in that area.

3 The Wishing Tree

[1] Edwin Wilson (ed.), Shirley Colless (research assistant), *Discovering the Domain*, Hale & Iremonger, Sydney 1986, pp. 30–33.
[2] Watkin Tench, *A Narrative of the Expedition to Botany Bay with an account of New South Wales, its productions, inhabitants, etc.*, J. Debrett, London 1789, pp. 118, 121, 122.
[3] Patrick White, *Voss*, Penguin, Harmondsworth 1963, p. 34.
[4] BG/AR, 1902, as updated and reprinted in Wilson, *Discovering the Domain.*
[5] J. H. Maiden, Botanic Gardens and Domain Report 1897 (25 March 1898).
[6] Walter W. Froggatt (ed.) from Notes of the late J. H. Maiden, *JRAHS*, Vol. XVII, Part III, 1931, pp. 152, 153.
[7] R. H. Cambage (ed.) from Notes of the late J. H. Maiden, 'History of the Sydney Botanic Gardens', *JRAHS*, Vol. XIV, Part I, 1928, p. 26.
[8] J. H. Maiden, *Royal Botanic Gardens Kew Bulletin*, London, No. 6, 1906, p. 208.
[9] Roy H. Goddard, *The Life and Times of James Milson*, Georgian House, Melbourne 1955, p. 35.
[10] Froggatt, *op. cit.*, p. 148.
[11] *Sydney Morning Herald*, 26 June 1846.
[12] *Ibid.*
[13] Froggatt, *op. cit.*, p. 8.
[14] Froggatt, *op. cit.*, p. 148.
[15] *Ibid.*
[16] *Ibid*; *Sydney Morning Herald*, 30 September 1897.
[17] Lionel Gilbert, *The Royal Botanic Gardens, Sydney: A History 1816–1985*, Oxford University Press, Melbourne, 1986, p. 26 — from J. H. Maiden in *Kew Bulletin*, 1906, p. 209 (from *Sydney Morning Herald*, 21 April 1906 — see also *Sydney Morning Herald*, 16, 23 June 1917).
[18] Ethel Turner, *The Family at Misrule*, Ward Lock & Co., London 1895, p. 120.
[19] Edwin Wilson, *Banyan*, Woodbine Press, Sydney 1982, p. 31. The martyr referred to in the poem is Joseph Gerrald, see p. 127).
[20] R. H. Anderson, *An ABC of the Royal Botanic Gardens, Sydney*, VCN Blight, Government Printer, Sydney 1965, p. 65. On 10 March 1825 Governor Brisbane records: 'I have added five acres to the Old Garden' (west of the creek in the Middle Garden). On 28 January 1828 Charles Fraser, writing to the Colonial Secretary, states 'Here the indigenous trees, collected on the Brisbane river by Allan Cunningham and Charles Fraser were planted'. From JRAHS, Vol. XVII, Part II, 1931, p. 130. Froggatt, *op. cit.*, p. 150, confirms that both Fraser and Cunningham accompanied John Oxley on his expeditions. Fraser's 'Journal of a Two Months' Residence on the Banks of the Rivers Brisbane and Logan, on the east Coast of New Holland' was published by W. J. Hooker in his *Botanical Miscellany*, Volume 1, p. 238. On this expedition Charles Fraser had been directed, amongst other things, to establish a public garden at Brisbane Town. Quotations from above come from pp. 240, 250, and 269 respectively.
[21] State Archives, Supplement No. 9.
[22] Kew Correspondence, Kew Collectors, Vols. 6, 7; Allan Cunningham.
[23] Froggatt, *op. cit.*, p. 150.
[24] *Sydney Gazette*, 20 May 1830.
[25] Gilbert, *op. cit.*, p. 49.
[26] Froggatt, *op. cit.*, p. 150.
[27] Maiden, *op. cit.*, p. 211.
[28] Wilson, *Discovering the Domain*, pp. 14–15; from a facsimile edition, *An Illustrated Guide to Sydney 1882* (Gibbs Shallard & Co.), Angus & Robertson, Sydney 1982 p. 113.
[29] Report of Charles Moore to Legislative Assembly 1870–71 (State Archives, Supplement No. 9).

[30] Research by Betty Jacobs, from reports of Charles Moore, 10 June 1856 and 24 May 1871; William Foster, Vice-President of the Royal Australian Historical Society, in some notes on our History File RBG.208, said 1956 was the centenary of the 'Azalea Garden'; Report by Charles Moore to Legislative Assembly 1870–71 (State Archives, Supplement No. 9).
[31] Charles Davis, 'A Stroll through the Botanical Gardens', *Sydney Quarterly Magazine*, September 1888, p. 211.
[32] Report by Charles Moore to Legislative Assembly 1870–71 (State Archives, Supplement No. 9).
[33] John Dalton, *The Cruise of Her Majesty's Ship 'Bacchante' 1879–1882*, McMillan & Co., London 1886, p. 584.
[34] Alan Gill, 'Little Princes who Captivated Australia', *Sydney Morning Herald*, 29 July 1981.
[35] *Ibid.*
[36] *Ibid.*
[37] *Ibid.*
[38] *Ibid.*
[39] *Ibid.*
[40] BG/AR, 1902, p. 4.
[41] Probable identification of this tree (from an earlier publication by J. H. Maiden) was made by Tony Rodd in 1981, prior to the publication of *Royal Botanic Gardens Sydney* in 1982.
[42] *Sun-Herald*, 25 August 1985.
[43] *Sydney Morning Herald*, 21 November 1988.
[44] Edwin Wilson, *The Rose Garden*, Woodbine Press, Sydney 1991, p. 75.
[45] Anderson, *op. cit.*, p. 66.
[46] *Ibid.* Additional information supplied by George Chippendale.
[47] Anderson, *op. cit.*, p. 66.
[48] Dr Phillip Norrie, *Vineyards of Sydney, Cradle of the Australian Wine Industry, from First Settlement to Today*, Horwitz Grahame, Sydney 1990, p. 19.
[49] Told to me by Ian McLellan, December 1988.
[50] F. A. Roach, *Cultivated Fruits of Britain: Their Origin and History*, Basil Blackwell, Oxford 1985, p. 215.
[51] *Ibid.*, p. 216.
[52] Sidney Beisly, *Shakespeare's Garden: The Plants and Flowers named in Shakespeare's works*, Longman, Green, Longman, Robert and Green, London 1864, p. 132.
[53] Roach, *op. cit.*, p. 217.
[54] *Ibid.*, p. 218.
[55] The Reverend Doctor Thomas Campbell, *The Diary of a Visit to England by an Irishman*, Waugh & Cox, Sydney 1854, p. 106. (The Reverend Campbell travelled with Boswell and Johnson).
[56] Beisly, *op. cit.*, p. 131.
[57] Governor King, writing to Sir Joseph Banks on 20 November 1800: 'There are a few Cape plants, such as vine cuttings and a mulberry or two'. From *JRAHS*, Vol. XVII, Part II, 1931, p. 140.
[58] Poem by Edwin Wilson, written 30 October 1990 after eating a fruit from 'Shakespeare's Mulberry tree' on Lawn 17, and published in *The Rose Garden*, *op. cit.*, p. 83.
[59] Edwin Wilson, 'The Rose' — a variant form first published in 1988 in The Rose Garden leaflet, Royal Botanic Gardens Sydney, and *The Rose Garden, op. cit.*, p. 15. This poem was first drafted during a talk by Professor Carrick Chambers to Senior Citizens in the Rose Garden Pavilion on 17 March 1988.

4 The Public Works

[1] David Churches, *The Cultural Significance of the Royal Botanic Gardens*, Sydney, 1990, p. 157.
[2] BG/AR, 1909, p. 30.
[3] Walter W. Froggatt (ed.) from Notes of the late J. H. Maiden, *JRAHS*, Vol. XVII, Part III, 1931, p. 154.
[4] *Ibid.*, p. 155.
[5] *Ibid.*
[6] *Ibid.*
[7] *Ibid.*, p. 156.
[8] *Ibid.*
[9] E. M. (Elsie) Webster, *Whirlwinds in the Plain, Ludwig Leichhardt — Friends, Foes and History*, Melbourne University Press, Melbourne 1980, p. 52.
[10] *Ibid.*, p. 75. Most of the seed collected from the previous expedition donated to the Gardens had languished and died. 'The new (temporary) superintendent of the Botanic Garden, inspecting some of Leichhardt's earlier finds at Camden, had dismissed them as "all poor and miserable".' *Ibid.*, p. 222.
[11] BG/AR, 1856, p. 2.
[12] BG/AR, 1854, p. 2.
[13] BG/AR, 1853, p. 2.
[14] BG/AR, 1858, p. 2.
[15] State Archives, Supplement No. 9.
[16] Report by Charles Moore to Legislative Assembly 1870–71, State Archives, Supplement No. 9.
[17] Froggatt, *op. cit.*, p. 156.
[18] Information supplied by Karen Wilson (Royal Botanic Gardens, Sydney).
[19] Webster, *op. cit.*, p. 272.
[20] *Ibid.*, p. 308.
[21] Information supplied by Karen Wilson (Royal Botanic Gardens, Sydney).
[22] Froggatt, *op. cit.*, p. 157.
[23] *Town and Country Journal*, 6 April 1878.
[24] Froggatt, *op. cit.*, p. 156.
[25] J. Broadbent report, May 1976, for conservation recommendations.
[26] BG/AR, 1899, p. 4.
[27] T. S. Champion, Evolution of Conservatorium of Music, Sydney, *Royal Australian Historical Society*, Vol. 33, pp. 313–316.
[28] Daphne Robilliard has shown me an old photograph of her great grandfather, James McFarland, with her grandmother, Mary Jane McFarland (smaller child in photograph), outside a two-storeyed stone cottage in the Gardens (then called Cottage 1, in the region of Cottage 4).
[29] Told to me by Eva Levis, niece of Edward Ward, who lived with him in the Gardens when completing high school, 1928–33.
[30] BG/AR, 1921, p. 2.
[31] Churches, *op. cit.*, p. 156.
[32] Research by Robert Staas, Noel Bell, Ridley Smith & Partners Pty Ltd.
[33] *Ibid.*
[34] *Ibid.*
[35] *Ibid.*
[36] BG/AR, 1908, p. 25.
[37] Research by Robert Staas.

[38] Information on glasshouses in Depot from the 1978 plan — research by David Churches, *op. cit.*, 9.19.
[39] Churches, *op. cit.*, p. 578.
[40] BG/AR, 1899, p. 3.
[41] David Churches, *op. cit.*, p. 97 (Mr Hugh Dixson is not to be confused with Mr William Dixson of the Dixson Wing of the State Library of New South Wales).
[42] BG/AR, 1899, p. 3.
[43] BG/AR, 1900, p. 3.
[44] BG/AR, 1909, p. 3.
[45] BG/AR, 1911, p. 4.
[46] BG/AR, 1923, p. 1.
[47] J. H. Maiden, *Guide to the Botanic Gardens*, Government Printer, Sydney 1903, p. 19.
[48] W. R. Watson and A. N. Rodd, Royal Botanic Gardens, Leaflet, 'The Glasshouse'.
[49] Research by Robert Staas.
[50] Churches, *op. cit.*, p. 660.
[51] *Ibid*, PWD/AR, 1897.
[52] *Ibid*, BG/AR, 1901, p. 27.
[53] *Ibid*, BG/ARs, 1906, p. 5; 1908, p. 3; 1911, p. 4; 1915, p. 5.
[54] Sumner Locke Elliott, *Water Under the Bridge*, Macmillan, Melbourne 1977, p. 72.
[55] BG/AR, 1878/79, p. 4.
[56] *Town and Country Journal*, 6 April 1878.
[57] Told to me on 15 August 1989 by Jim Stovold, grandson of (old) Jim Stovold.
[58] Churches, *op. cit.*, p. 179.
[59] Compiled from Edwin Wilson (ed.), Shirley Colless (research assistant), *Discovering the Domain*, Hale & Iremonger, Sydney 1986; and the conservation report by J. Broadbent, May 1976.
[60] Churches, *op. cit.*, p. 818.
[61] *Ibid.*
[62] Told to me by Ms Oi Choong, Public Works Department, as read in an as yet unlocated article published in *Sydney Morning Herald*.
[63] Provided for the Director of Gardens by David Churches, Architect, Special Projects Branch, Government Architect's Office.
[64] BG/AR, 1897.
[65] Description by J. Broadbent for proposed heritage listing by the National Trust, May 1976.
[65] BG/AR, 1897, p. 17.
[67] BG/AR, 1901, p. 28.
[68] BG/AR, 1902; BG/AR, 1897, p. 17.
[69] BG/AR, 1907, p. 27.
[70] BG/AR, 1919, p. 13.
[71] Separate note provided for Professor Carrick Chambers, Director of Gardens, by David Churches.
[72] *Ibid.*
[73] *Sydney Morning Herald*, 26 June 1846.
[74] Information supplied by Mrs Joy Thompson (Royal Botanic Gardens, Sydney).
[75] BG/AR, 1852, p. 3.
[76] BG/AR, 1910, p. 27.
[77] Churches, *op. cit.*, p. 631.
[78] *Ibid*; BG/AR, 1914, p. 4.
[79] Churches, *op. cit.*, p. 631.
[80] BG/AR, 1899, p. 24.
[81] BG/AR, 1905, p. 26.
[82] BG/AR, 1911, p. 26.
[83] BG/AR, 1899, p. 4.
[84] David Churches, *Historic Buildings in Sydney*, Public Works Department and Heritage Council of New South Wales, No. 13; National Trust Listing of Buildings, prepared in May 1976 by James Broadbent.
[85] State Archives, Supplement No. 9.
[86] Edwin Wilson (ed.), Shirley Colless (research assistant), *Discovering the Domain*, Hale & Iremonger, Sydney 1986, p. 5.
[87] BG/AR 1908, p. 3.

5 Arcadia Revisited

[1] From Tom Hillard's compilation 'The Classical Tradition and Classical Memories in the Royal Botanic Gardens', as a companion to Nola Thompson's 'Classical Tradition in the Botanic Garden' for the Macquarie Ancient History Association's peripateia and picnic in the Gardens, 23 March 1991.
[2] Kathleen Wall, Volunteer Guide, 'Renaissance Lady' and former science lecturer at teachers' colleges, assisted me with this and other manuscripts over the years and has helped to make me less of a chauvinist (in the modern sense of the word) by her intellect, chastisement and example.
[3] RBG File 7(A), letter dated 23 May 1971 from Mr H. Dominy to the Director, Royal Botanical (*sic*) Gardens.
[4] *Ibid.*, reply to Mr H. Dominy by Dr J. S. Beard, Director, Royal Botanic Gardens, 31 May 1971.
[5] Hillard, *op. cit.*
[6] A. J. Clark, *Griffin* '77, magazine of Hurstville Boys' High School, p. 5.
[7] Research by Betty Jacobs.
[8] Clark, *op. cit.*
[9] Information compiled from RBG File 7(A).
[10] Hillard, *op. cit.*
[11] *Ibid.*
[12] Edwin Wilson, *Songs of the Forest*, Hale & Iremonger, Sydney 1990, p. 44. When I was a young man the beer-garden at the Coolangatta Hotel was called 'The Cabbage Patch', because the Victorian tourists thought its staghorn ferns were cabbages. Of the 30 species of *Acanthus*, *A. spinosus* (and not *A. mollis*) is believed to have inspired the leaf motif on Corinthian columns.
[13] Information supplied by Jacqueline Voignier-Marshall of the Australian Byron Society.
[14] Hillard, *op. cit.*
[15] The philosopher Demosthenes is reputed to have overcome a stutter and honed his orating skills by addressing the ocean waves with a mouth full of stones.
[16] *National Art Gallery of New South Wales Illustrated Catalogue*, Sydney 1926, p. 203.
[17] J. H. Maiden, *Royal Botanic Gardens Kew Bulletin*, No. 6, 1906, p. 206.
[18] *Sydney Morning Herald*, 29 January 1838.
[19] *Sydney Morning Herald*, 26 June 1846.
[20] *Ibid.*
[21] BG/AR, 1902, p. 4.
[22] Maiden, *op. cit.*, p. 216.
[23] Maiden, *op. cit.*, p. 206.

[24] Information supplied by K.L. Zywuszko, sculptor and conservator, Melbourne.
[25] BG/AR, 1918, p. 3.
[26] BG/AR, 1923. p. 1.
[27] David Engel, *Japanese Gardens for Today*; from information given to me by Diane Alford.
[28] Translation by Yasuko Claremont.
[29] Words (and music) of cantata, written by Henry Kendall, were supplied by Ian Watchorn, Curator of Music, Museum of Applied Arts and Sciences (Powerhouse).
[30] Renée Free, 'Late Victorian, Edwardian, and French Sculptures', *Quarterly*, January 1972, p. 651.
[31] R. H. Anderson, *An ABC of the Royal Botanic Gardens, Sydney*, V. C. N. Blight, Government Printer, Sydney 1965, p. 51.
[32] Information supplied by Mr R. S. Dickson, great-grandson of Edward Sanders.
[33] BG/AR, 1898, p. 17.
[34] RBG File 7(A) Minute K. Mair to Director-General of Agriculture, 15 December 1966.
[35] Letter 8 January 1991 from Dominique Daire, great-grandnephew of Arthur le Duc.
[36] Centennial Park File Number, FN63.
[37] Avenel Mitchell *et al.*, Art Gallery of New South Wales, Australian Sculpture, 1890–1919, Project Exhibition, Court 9, 22 January-22 February 1987.
[38] BG/AR, 1902, p. 31.
[39] BG/AR, 1899, p. 35.
[40] Research by Judith Hunter: *Sydney Mail*, 27 September 1879, p. 515; *Town and Country Journal*, 27 September 1879, p. 582.
[41] Research by Judith Hunter, official record of the International Exhibition Sydney, 1879–80, ML 606/S RL MK1S1, p. 506.
[42] *Ibid.*
[43] David Churches, *The Cultural Significance of the Royal Botanic Gardens, Sydney*, 1990, draft manuscript, 10.1.
[44] Research by Judith Hunter, *Illustrated Sydney* News, 25 October 1882, p. 6.
[45] J. H. Maiden, *Guide to the Botanic Gardens*, Government Printer, Sydney 1903, p. 4.
[46] *Ibid.*
[47] Research by Tony Jeans: *Sydney Morning Herald*, 20 January 1883.
[48] Letter 141, Sir Henry Parkes correspondence, Vol. 39, Q709.901/M.
[49] BG/AR, 1905, p. 33.
[50] Harry Craig, grandson of J. H. Maiden, told me his grandfather used to keep the notes for his various books and publications in brown paper bags and stored them in shoe boxes.
[51] J. C. L. Fitzpatrick, *The Good Old Days* Cumberland Argus, Parramatta 1910, pp. 140–141, quoting from the *Australasian Sketcher* (Windsor), with a follow-up reference in a memorial edition of *The Advocate* (Hornsby) on 16 July 1969, p. 43.
[52] *Sydney Morning Herald*, 19 January 1898.
[53] *Illawarra Mercury*, 24 March 1922.
[54] Helen Vellacott (ed.), *Some Recollections of a Happy Life: Marianne North in Australia and New Zealand*, Edward Arnold Australia, Melbourne 1986, p. 41.
[55] *Ibid.*, p. 40. According to the research of Janet Dawbin, Volunteer Guide, Royal Botanic Gardens, *Ficus bennettii* (now *Ficus habrophylia*) from the New Hebrides — Bennett had visited the New Hebrides in his youth — once grew in Lawn 5, Bed 15c, and Bed 23, but none is here now. J. H. Maiden mentions *Ficus moorii* in 1908 but it has since disappeared from the Gardens.
[56] *National Art Gallery of New South Wales Illustrated Catalogue*, Sydney 1926, p. 202. RBG File 7(A); the letter from Job Hanson to Sir Henry Parkes quoted earlier implies this statue may have been done by Canova.

6 The Plaque near the Ladies Toilet

[1] BG/AR, 1965 (as part of N.S.W. Department of Agriculture Report), p. 100.
[2] RBG File 7(A) Minute R. H. Anderson to Director-General of Agriculture.
[3] Abraham Rees, *The Cyclopaedia or Universal Dictionary of Arts, Sciences and Literature, Paternoster Row, Longman, Hurst, Orme Brown, 1819*, Vol. X; James Mackay, *The Dictionary of Western Sculptors in Bronze*, Woodbridge, Suffolk, Antique Collectors' Club, 1977, p. 83; translation by Dr L. A. S. Johnson.
[4] Frank Crowley, (ed.), *A Documentary History of Australia*, Vol. 2, *1841–74*, Nelson, Melbourne 1973–1980, p. 25.
[5] *Sydney Morning Herald*, 20 June 1846.
[6] Despatches from Lord John Russell to Sir George Gipps 1839. (Note 10, pp. 68, 69 & 373), p. 850.
[7] Meredith, *op. cit.*, p. 41.
[8] Philip Geeves, *Philip Geeves' Sydney*, Angus & Robertson, Sydney 1981, p. 58; *Sydney Morning Herald*, 20 June 1846; also Leigh & Co's *Handbook to Sydney and Suburbs, 1867*, p. 10; Mrs Charles Meredith, *Notes and Sketches of New South Wales*, p. 41, had indicated that she thought the statue was executed by the English sculptor Westmacott.
[9] Despatches from Lord John Russell to Sir George Gipps, 1841, pp. 298, 373–374.
[10] *Ibid.* (Resolutions, 9 February 1839), p. 70.
[11] *Ibid.*
[12] *Ibid.* (Note 11, p. 69), p. 850.
[13] Leigh & Co's *Handbook to Sydney and Suburbs, 1867*, p. 22.
[14] *Sydney Morning Herald*, 12 April 1842.
[15] Edwin Wilson (ed.) Shirley Colless (research assistant), *Discovering the Domain*, Hale & Iremonger, Sydney 1986, p. 14.
[16] Research by Shirley Colless, article by Colonel Hubert Foster, RE: *Sydney Morning Herald* 17 January 1914; ML Newspaper cuttings, Q991.N, No. 35, p. 96.
[17] *Ibid.*
[18] During the October attack by the Russian general Menshikov upon Balaclava, the English HQ, the Heavy Brigade won a notable success over the Russian Cavalry, and later in the day the Light Brigade made its epic charge through the northern valley.
[19] *Sydney Morning Herald*, 17 January 1914.
[20] Beryl Henderson (ed.), *Monuments and Memorials*, The Royal Australian Historical Society Register, Sydney 1988, p. 178.
[21] Geeves, *op. cit.*, p. 48.
[22] Henderson, *op. cit.*, p. 177.
[23] Renée Free, *Art Gallery of New South Wales Quarterly*, Vol. 13, No. 2, January 1972, p. 651.
[24] *Sydney Mail*, 6 June 1923, carried a full-page photograph and text, entitled 'A Noble Vista'.
[25] BG/AR, 1909, p. 29.

[26] BG/AR, 1909, p. 29.
[27] Henderson, *op. cit.*, p. 177.
[28] David J. Jones, *A Source of Inspiration and Delight. The Buildings of the State Library of New South Wales Since 1826*, Library Council of N.S.W., Sydney 1988, p. 162.
[29] Edmund Campion reviewing *Henry Lawson: A Life*, by Colin Roderick, *The Bulletin*, 21 May 1991, p. 122.
[30] *Labor Daily*, 27 August 1930.
[31] RBG File 7(A) — letter from Mr G. R. Crawford, Minister for Agriculture, to G. F. Freudenstein, MLA, Assistant Treasurer and Minister for Cultural Affairs, dated 27 December 1971.
[32] D. Sheedy, submission for Heritage Recommendations, May 1976.
[33] Told to me by Dr Glen Hunt, The Australian Museum. This story was told to him by his mother.
[34] Campion, *op. cit.*, p. 112.
[35] Edwin Wilson, *Banyan*, Woodbine Press, Sydney 1982, p. 58.
[36] Geeves, *op. cit.*, p. 94.
[37] DA/AR, 1967, p. 119.
[38] Henderson, *op. cit.*, p. 178.
[39] Art Gallery of New South Wales Sculpture Catalogue.
[40] Art Gallery of New South Wales British Sculpture Catalogue.
[41] BG/AR, 1918, p. 12.
[42] BG/AR, 1918, p. 12.
[43] BG/AR, 1920, p. 11.
[44] BG/AR, 1924 (for 1922), p. 2.
[45] D. Sheedy, report in May 1976 for proposed listing on a register of the National Estate.
[46] Wilson, *Discovering the Domain*, p. 24.
[47] Olive Lawson, *The First Voice of Australian Feminism: Excerpts from Louisa Lawson's The Dawn*, 1888–1895, Simon Schuster, Sydney 1990, pp. 11–12.
[48] BG/AR, 1904.
[49] BG/AR, 1912.
[50] N. K. Lamport (a relative of John Robertson), speaking to Edwin Wilson on 28 April 1982.
[51] Renée Free, 'The Art of Frank and Margel Hinder 1930–1980', *Art and Australia*, Vol. 18, No. 3, Autumn 1981, p. 247.

7 Dearly Departed

[1] Alan Gill, 'Little Princes Who Captivated Australia', *Sydney Morning Herald*, 29 July 1981.
[1] Philip Geeves, *Philip Geeves' Sydney*, Angus & Robertson, Sydney 1981, p. 116.
[3] Geeves, *op. cit.*, p. 116.
[4] *Ibid.*, p. 134.
[5] Research by Betty Jacobs.
[6] Letter from F. W. (Frank) Gardiner (grandfather of Joy Thompson, Royal Botanic Gardens, Sydney) to his father, 16 April 1866.
[7] Beryl Henderson (ed.), *Monuments and Memorials*, The Royal Australian Historical Society Register, Sydney 1988, p. 177.
[8] Geeves, *op. cit.*, p. 134.
[9] *Ibid.*, p. 116.
[10] *Ibid.*
[11] Ure Smith Production, *Portrait of Sydney: 1950s*, Ure Smith Pty Ltd, Sydney *c.* 1953, p. 25.
[12] Geeves, *op. cit.*, p. 134.
[13] *Sun*, 23 May 1930.
[14] BG/AR, 1901, pp. 4–5.
[15] Walter W. Froggatt (ed.) from Notes of the late J. H. Maiden, *JRAHS*, Vol. XVII, Part III, 1931, p. 160.
[16] Interview with ex-employee Sid Morrison by Heather Edwards 26 June 1990. It was one of his jobs to ring the bell which was still being used when he retired in 1958.
[17] Froggatt, *op. cit.*, p. 205.
[18] BG/AR, 1906, p. 33.
[19] *Sydney Morning Herald*, 22 February 1860.
[20] Information supplied by Enid Buckley, 25 November 1991, daughter of Hannah Fitzsimons.
[21] BG/AR, 1908, p. 25.
[22] Charles Davis, 'A Stroll through the Botanical Gardens', *Sydney Quarterly Magazine*, September 1888, p. 210.
[23] *The Australian Encyclopaedia*, 4th edn, Grolier Society of Australia, Sydney 1983, Vol. 2, p. 130.
[24] Froggatt, *op. cit.*, p. 158.
[25] *Ibid.*
[26] *Ibid.*
[27] David Churches, *The Cultural Significance of the Royal Botanic Gardens, Sydney*, 1990, p. 791.
[28] BG/AR, 1915, p. 40.
[29] Walter W. Froggatt (ed.) from Notes of the late J. H. Maiden, *JRAHS*, Vol. XVII, 1931, Part III, p. 158.
[30] Alan McCulloch, *Encyclopaedia of Australian Art*, Hawthorn Victoria, Hutchinson, 1984, Volume 1, p. 639. Information given to me on 1 June 1990 by Mrs June Parkins, great granddaughter of William Jones.
[31] J. P. McGuanne, 'Bennilong [*sic*] Point and Fort Macquarie', *JRAHS*, Vol. 16, 1/2, 1901, pp. 9–13.
[32] *Ibid.*, p. 10.
[33] Leigh & Co's *Handbook to Sydney and Suburbs, 1867*, p. 24; RBG File 208.
[34] Information provided in letter from Roy H. E. Harvey of the Royal Australian Artillery Historical Society, 26 November 1986.
[35] Alan Sharp (comp.), *Colonial New South Wales, 1853–1894*, Harper & Row, Sydney 1979, p. 89.
[36] *Sydney Mail*, 3 July 1929.
[37] Leigh & Co's *Handbook to Sydney and Suburbs, 1867*, p. 23.
[38] Churches, *op. cit.*, p. 792.
[39] J. H. Maiden, *Guide to the Botanic Gardens*, Government Printer, Sydney 1903, p. 76.
[40] BG/AR, 1899, p. 35.
[41] Geeves, *op. cit.*, p. 123.
[42] Lionel Gilbert, *The Royal Botanic Gardens, Sydney: A History 1816–1985*, Oxford University Press, Melbourne 1986, p. 184; illustration from J. W. Waugh, *The Strangers Guide to Sydney, 1861.*
[43] From *The Political Martyrs of Scotland Persecuted During the Years 1793 and 1794*, Steam Press of Peter Brown Printer, 19 St James' Square, Edinburgh (no date).
[44] *The Defence of Joseph Gerrald on a Charge of Sedition*, etc., printed for J. Ridgway, No. 1 York Street, St James' Square, Edinburgh 1794.
[45] BG/AR, 1905, p. 33.
[46] Churches, *op. cit.*, p. 788; supplementary photographs, BG/AR, 1914.

[47] BG/AR, 1909, p. 29.
[48] As published in Gilbert, *op. cit.*, p. 184; from J. W. Waugh, *The Strangers Guide to Sydney*, 1861.
[49] R. S. Horan, *Fort Street — The School*, Honeysett Publications Pty Limited, Sydney 1989, p. 250; photo of statue in *Fortian*, 1976, p. 7.
[50] *National Art Gallery of New South Wales Illustrated Catalogue*, Sydney 1926, p. 203.
[51] *Ibid*, p. 208.
[52] R. H. Anderson, *An ABC of the Royal Botanic Gardens, Sydney*, V. C. N. Blight, Government Printer, Sydney 1965, p. 50.
[53] Letter from Dr Ulrike Gauss, Assistant Keeper, Staatsgalerie Stuttgart, to Renée Free, Curator of European Art, Art Gallery of New South Wales, 16 February 1973, RBG File 7(B).
[54] Information supplied by Alan Tillsley (Royal Botanic Gardens, Sydney).
[55] *Sydney Morning Herald*, 10 January 1883.
[56] *Art Journal*, 1894, p. 199.
[57] Renée Free letter, RBG File 7(B).
[58] RBG File 7(A).
[59] RBG File 7(A).
[60] Nan McDonald, *Selected Poems*, Angus & Robertson, Sydney 1969, p. 25.
[61] Edwin Wilson, *The Rose Garden*, Woodbine Press, Sydney 1991, p. 23.
[62] Anderson, *op. cit.*, p. 52.
[63] *Sydney Morning Herald*, 'News of the Day' column, 10 January 1883.
[64] Froggatt, *op. cit.*, p. 160.
[65] Maiden, *Guide to the Botanic Gardens*, p. 4.
[66] BG/AR, 1909, p. 29.
[67] Froggatt, *op. cit.*, p. 160.
[68] *Sun*, 23 May 1930.
[69] *Smith's Weekly*, 24 May 1930.
[70] Told to me originally by Alan Tillsley, Royal Botanic Gardens, Sydney. Supported by article in the *Sydney Morning Herald*, 26 June 1846.
[71] RBG File 7(A).
[72] *National Art Gallery of New South Wales, Illustrated Catalogue*, Sydney 1926, p. 206.

8 Centennial Park

[1] Letter from Sir Henry Parkes to Charles Moore, State Archives, Supplement No. 9.
[2] H. Furniss, *Australian Sketches*, Ward Lock, London 1899, p. 182.
[3] A. J. Clark, *Griffin '77*, Hurstville Boys High School, 1977, p. 5.
[4] Told to me by John Mortimer, Director, Centennial Park, 8 December 1989. He was not able to locate the source of this information. Additional Centennial Park material subsequently published in the *Centennial Park Heritage Study*, PWD Report No. 90036.
[5] Centennial Park reference 89/347.
[6] Information supplied by John Mortimer.
[7] *Sydney Morning Herald*, 13 November 1974.
[8] BG/AR, 1899, p. 27.
[9] BG/AR, 1900, p. 36.
[10] Centennial Park reference FN62B.
[11] A. J. Clark, *Griffin '77*, Hurstville Boys High School, 1977, p. 5.
[12] *Ibid.*
[13] *The Australian*, 31 December 1971.
[14] Paul Ashton and Kate Blackmore, *Centennial Park: A History*, Centennial Park Trust, New South Wales University Press, Kensington 1988, p. 40.
[15] *Sydney Morning Herald*, 30 December 1971; *The Australian*, 31 December 1971; *Daily Mirror*, 30 December 1971.
[16] *Daily Mirror*, 25 January 1972.
[17] *Ibid*, 25 January 1972.
[18] *Daily Telegraph*, 24 February 1972.
[19] *Daily Telegraph*, 24 February 1972.
[20] Letter from Dr J. S. Beard to Mrs D. State, Honorary Secretary of the Sydney Branch of the Dickens Fellowship, 3 May 1972.
[21] RBG File 7(B).
[22] *Sydney Morning Herald*, 13 November 1974.
[23] *Sydney Morning Herald*, 13 November 1974.
[24] Letter from Ian Stubbin on RBG 7(B) File, 19 October 1975.
[25] BG/AR, 1902, p. 25. J. H. Maiden refers to the availability of some old Doric columns from Darlinghurst Court House. There is considerable confusion about the source of these columns — Centennial Park literature refers to both the 'Museum' and the (original) Sydney Post Office (RBG File 7(B)). These may all be 'correct' in so far as the peripatetic 'Museum' (The Australian Museum) was once in a shed attached to the Bent Street residence of the Judge Advocate (previous site of the Post Office) and later at the Court House at Darlinghurst from 1841–47 (information from Ronald Strahan (ed.), *Rare and Curious Specimens: A History of The Australian Museum*, Australian Museum, Sydney 1979, p. 18).
[26] Translated 30 November 1990 by Irma Trunov, Royal Botanic Gardens, Sydney.
[27] Centennial Park reference FN61A.
[28] BG/AR, 1912, p. 29.
[29] BG/AR, 1898, p. 20.
[30] *Sydney Morning Herald*, 2 January 1901.
[31] The Federation Pavilion, Centennial Park Leaflet.
[32] BG/AR, 1905, p. 28.
[33] BG/AR, 1905, p. 28.
[34] Minute from L. A. S. Johnson, Director, 13 November 1973, regarding letter from Lieutenant-Commander W. M. Swan (retired) regarding the dates on the Federation Monument. Additional information from J. S. Battye, *Western Australia: A History from its Discovery to the Inauguration of the Commonwealth*, Oxford, England 1924, p. 71.
[35] The Federation Pavilion, Centennial Park Leaflet.
[36] *Ibid.*
[37] BG/AR, 1901, p. 27.
[38] Information provided by Peter Roberts, Gould League.
[39] BG/AR, 1898, p. 20.
[40] Philip Geeves, *Philip Geeves' Sydney*, Angus & Robertson, Sydney 1981, p. 52.
[41] Graeme Sturgeon, *The Development of Australian Sculpture, 1788–1975*, Thames St. London, Hudson 1978, p. 33.
[42] Letter to Sir Henry Parkes by James Barnet, 3 April 1891.
[43] RBG File 7(B) — Minute L. A. S. Johnson, Director, to Director-General of Agriculture, 29 August 1972.

Royal Botanic Gardens Sydney and Domain

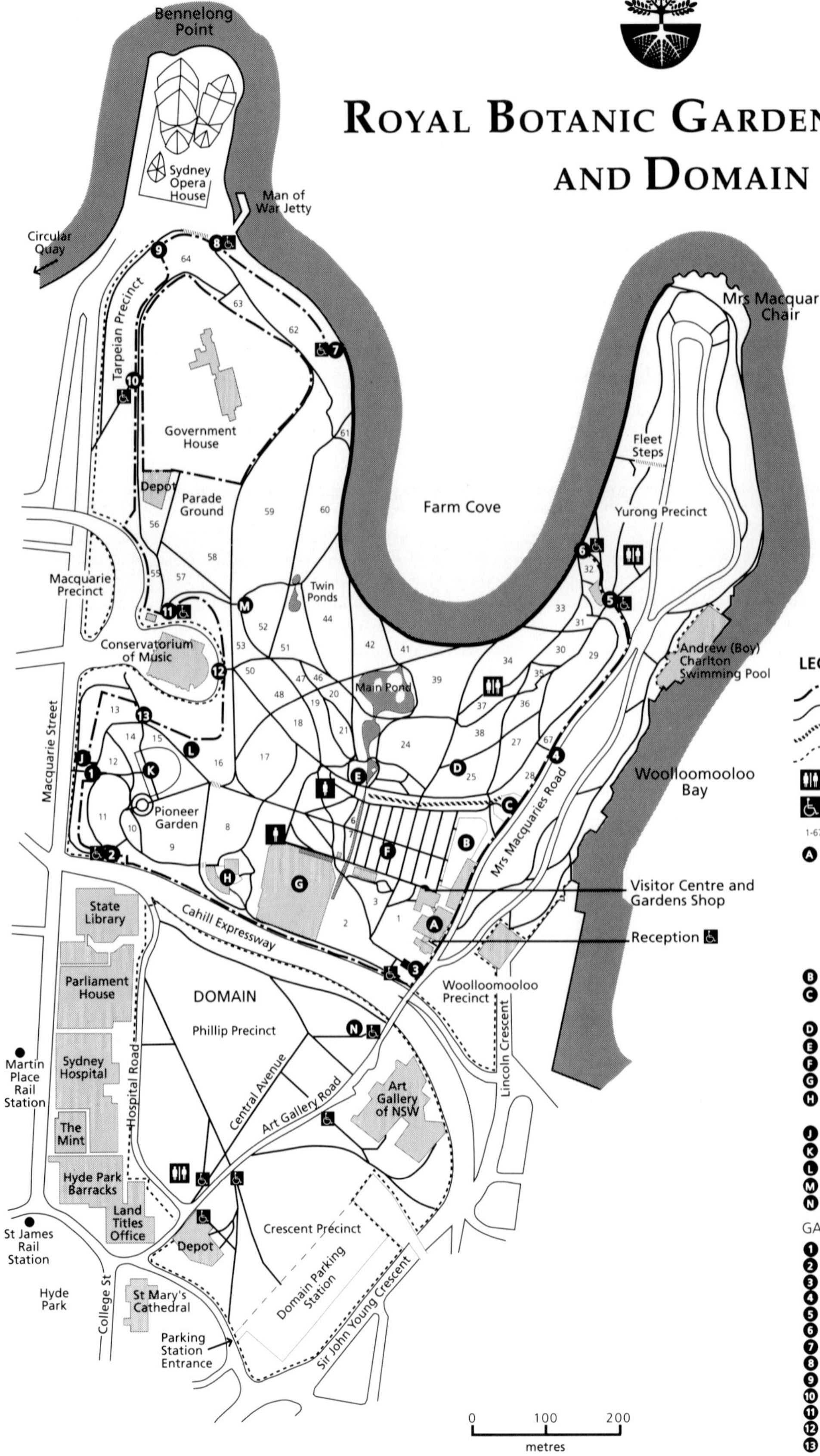

LEGEND

- Boundary of the Gardens
- Paths
- Macquarie Wall
- Boundary of the Domain
- Toilets
- Wheelchair access points
- 1-67 Lawns

A Royal Botanic Gardens Offices
- Reception & Administration
- Visitor Centre & Gardens Shop
- Maiden Theatre
- Education Unit
- National Herbarium of NSW
- Plant Identification Service

B Succulent Garden
C Friends of the Royal Botanic Gardens (Lion Gate Lodge)
D Maiden Pavilion
E Gardens Restaurant & Kiosk
F *First Farm* Exhibition
G Main Depot & Nursery
H Sydney Tropical Centre - the Arc & the Pyramid
J Palace Gate Kiosk
K Rose Garden & Pavilion
L Grass Plots
M Vista Pavilion
N Restaurant on the Park & Kiosk

GATES

1. Palace Garden Gate
2. Morshead Fountain Gate
3. Woolloomooloo Gate
4. Henry Lawson Gate
5. Victoria Lodge Gate
6. Yurong Gate
7. Queen Elizabeth II Gate
8. Opera House Gate
9. Tarpeian Way Gate
10. Government House Delivery Gate
11. Government House Gate
12. Herb Garden Gate
13. Rose Garden Gate

4/92

Index

Map 2. The Royal Botanic Gardens and Domain, 1991.